Forests to
Fight Poverty

CONTRIBUTORS

CHARLES BENBROOK is a natural resources and environmental management consultant and president of Benbrook Consulting Services, based in Sandpoint, Idaho.

JOYCE K. BERRY is the director of the Environment and Natural Resource Policy Institute, Colorado State University, Fort Collins.

JOHN C. GORDON is the Pinchot Professor of Forestry at Yale University School of Forestry and Environmental Studies, New Haven, Connecticut.

RUBÉN GUEVARA is the director general of the Tropical Agriculture Research and Higher Education Center (CATIE), based in Turrialba, Costa Rica.

CALESTOUS JUMA is the executive secretary of the Secretariat of the Convention on Biological Diversity, based in Montreal.

TAPANI OKSANEN is the vice president of INDUFOR Oy, based in Helsinki, and senior advisor to the UNDP's Programme on Forests.

DAVID PEARCE is a professor at the Centre for Social and Economic Research on the Global Environment at the University College of London.

NANCY PELUSO is a professor in the Department of Environmental Science, Policy, and Management in the College of Natural Resources at the University of California, Berkeley.

RALPH SCHMIDT is the director of the UNDP Programme on Forests (PROFOR).

JOHN SPEARS is a consultant on sustainable forest management for the Tropical Forest Foundation of Alexandria, Virginia.

Forests to Fight Poverty

Creating National Strategies

Edited by Ralph Schmidt,

Joyce K. Berry, and John C. Gordon

Based on an initiative of the UNDP

Yale University Press
New Haven & London

Designed by James J. Johnson and set in Melior type
by The Composing Room of Michigan, Inc.
Printed in the United States of America by Vail-Ballou Press,
Binghamton, New York, on acid-free, recycled paper.
The post-consumer waste content of the recycled paper is 10–15 percent.

Library of Congress Cataloging-in-Publication Data

Forests to fight poverty : creating national strategies / edited by
Ralph Schmidt, Joyce K. Berry, and John C. Gordon.
p.   cm.
Includes bibliographical references and index.
ISBN 0-300-07309-7 (cloth : alk. paper). — ISBN 0-300-07845-5
(pbk. : alk. paper)

1. Sustainable forestry—Developing countries.   2. Sustainable
development—Developing countries.   3. Rural development—
Developing countries.   4. Forest policy—Developing countries.
I. Schmidt, Ralph (Ralph C.)   II. Berry, Joyce K.   III. Gordon, J. C.
(John C.), 1939–   .
SD247.5.F68   1999
333.75′15′091724—dc21   98-53024

A catalogue record of this book is available from the British Library.

The paper in this book meets the guidelines for permanence and
durability of the Committee on Production Guidelines for Book
Longevity of the Council on Library Resources.

10   9   8   7   6   5   4   3   2   1

Contents

Illustrations

Tables

Foreword

When forest programs are discussed, many themes are considered. Those working on programs in gender in development, sustainable food security and agriculture, health and nutrition, trade and sustainable development, water, biodiversity, energy, climate change, and land management should all be at the table. We want our foresters to consider all these dimensions as they focus on the sustainable management of forest lands for all their services and benefits.

These connections reflect the uses of the forests by the rural poor in developing countries. This is not a minor sector involving a small proportion of the society. Marginal agricultural lands where families practice subsistence farming in a shifting mosaic of farmland and forest are home to one-half of the people of Africa, one-third of the people of Asia, and one-quarter of the people of Latin America. The proportion of the poor in these areas is even greater.

Forests are intimately involved in these peoples' production and provision of food, water, energy, housing, and medicine. Subsistence farming families are using and managing forest vegetation every day to meet their most basic needs.

International and national organizations might learn much from the holistic, cross-sectoral approach practiced by rural people. We cannot allow the "forest agenda" to be fractionated into disparate programs.

Mirroring the local importance of forests is their global significance. The planetary ecosystem we are just beginning to understand—the biosphere—may have a minimum threshold of forest cover necessary to support a certain level of human habitation. When the dawn of the third millennium is considered from a historical perspective, some major trends will stand out. The doubling of the human population from five to ten billion in a few decades will be one of them. A precipitous drop in global forest cover will be another. We will have lost an area the size of Europe, including Scandinavia, in the last two decades of the twentieth century.

Forests exemplify many of the hard choices and conflicts of interest related to sustainable development. A long-term and cross-sectoral perspective is necessary in order to optimize human benefits, and there is always a tendency to rapidly liquidate natural capital that has accumulated slowly over long periods of time.

The United Nations Development Programme (UNDP), for example, most often works from a national perspective on issues related to sustainable human development (SHD). The loss of forests and the many development options they provide constitutes one of the most serious threats to SHD. The countries of Central America, western and southern Africa, and mainland Asia (with a couple of exceptions) have in the past few decades lost more than 90 percent of their closed forests. Of course Europe, excluding Russia and Scandinavia, lost a similar proportion of its forests long ago. Of countries once extensively forested, some have lost 99 percent of their closed forests; thirty have lost more than 90 percent. Of these, fourteen are least developed countries, and twelve have recently suffered violent disorder requiring UN peacekeeping forces. The list includes almost all countries where peacekeeping forces have recently been present. Extreme deforestation is a good indicator of the degradation of natural resources that is repeatedly associated with increasing civil strife and disorder.

For many reasons, then, this book takes the view that one cannot work on SHD for the poor in developing countries without integrating a concern for forests. A national vision of and strategy for forests needs to be firmly incorporated within an overall national development strategy. The fate of forests—and the many essentials they provide, especially for the poor—will depend on these national forest programs (nfps).

Analysis of these programs, often conducted by countries themselves, has shown that capacity-building is indispensable to their success. *Capacity-building* refers to efforts to assist countries in developing their ability to initiate, implement, and integrate sustainable development goals into the national fabric. Through strengthening policy and legislation, building infrastructure, training and educating people, transferring technology, and facilitating the equal participation of all groups in decision-making, the nation's power to perform its various functions is increased. A growing international consensus points to a capacity-building not so much based on centralized institutions and sophisticated training but related to innovative and challenging tasks such as participatory stakeholder involvement, cross-sectoral integration, and effective management of multiple funding sources. These are key bottlenecks that are not limited to forestry programs and are not by and large technical forestry matters.

We need good forestry and agriculture for successful nfps, but it will take more than just good forestry to make them successful. In order to sustain forests, the nfps need to be nested in an approach that provides SHD for the rural poor. Trade and macroeconomic policy, health, and education need close consideration, and this leads to comprehensive SHD programs.

Comprehensive and strengthened national programs to sustain forests must be based on two principles: (1) they must be country-driven and -initiated; and (2) they will, in many cases, require new and more effective international support using innovative means ranging across economic relations: macroeconomic policies, trade, and debt, for example. This mutual commitment would require new partnerships initiated and managed by countries themselves.

It can be convincingly argued that all major multilateral organizations are underfunding forest programs. It will take partnerships based on new equations of national commitment and international cooperation to generate the right level of support. In a recent meeting developing countries' nfp directors called for a new beginning by the international community to provide the needed support. Let us bring to the table the diversity necessary to provide that new beginning. This book will help that process.

JAMES GUSTAVE SPETH
Administrator of the UNDP

Acknowledgments

Following the landmark 1992 UN Conference on Environment and Development in Rio de Janeiro, intense discussions and efforts were launched within the United Nations Development Programme to pursue work in the sectoral areas most critical to an integrated vision of sustainable human development. Although the goal was clear, the exact means with which to achieve it were not. This book has evolved from that process as one part of our ongoing strategy.

This publication's production was supported financially and intellectually within the UNDP Bureau for Development Policy, Eimi Watanabe, director, and the Sustainable Energy and Environment Division, Roberto Lenton, director. Luis Gomez-Echeverri, then head of the Environment and Natural Resources program within UNDP and now director of its Public-Private Partnerships program, provided guidance on the overall concept of the book and on assembling the team. Editors and authors Ralph Schmidt, Joyce K. Berry, and John C. Gordon took this work from concept to outline and then manuscript. Joyce K. Berry took the lead on editing.

An extremely distinguished and busy set of individuals

was asked for contributions and graciously accepted. It is with deep gratitude that we would like to thank each of them. Their expertise is an asset not only to this publication but to the development community in general.

An editorial advisory group was composed of Mohamed Salleh Nor, of the Forest Research Institute in Malaysia, and authors Calestous Juma and Rubén Guevara, who provided much-appreciated direction on the outline and content of this book.

The chief administrator of the UNDP, James Gustave Speth, in addition to supplying the leadership that fosters innovation and creativity throughout the UNDP and especially within the Policy Bureau, has long displayed keen interest in and extensive knowledge of forest programs. Many of the ideas central to this book originated with him.

Stephen Dembner, the editor of the UN Food and Agriculture Organisation's forestry magazine, *Unasylva,* generously facilitated access to FAO's photographic library. We gratefully acknowledge that all photographs are courtesy of FAO.

Jean Thomson Black of Yale University Press collaborated with the editors most effectively and helpfully on issues of content, format, and presentation. Her support has been crucial to the success of this effort.

Susan Hochgraf provided graphic design for the charts, maps, and other figures. Namrita Kapur, Karol Maruska, and Cintra Agee contributed very able professional assistance to the completion of all the tasks that brought this manuscript to print.

Acronyms

ACTS	African Centre for Technology Studies
ASEAN	Association of Southeast Asian Nations
BS	Biodiversity Strategy (Convention on Biological Diversity)
CATIE	Tropical Agriculture Research and Higher Education Center (Latin America)
CBO	Community-based organization
CGIAR	Consultative Group for International Agricultural Research
CIFOR	Center for International Forestry Research
CS	Conservation Strategy (World Conservation Union—IUCN)
CSD	Commission on Sustainable Development
CVM	contingent valuation method
Dec.P	Program to Combat Desertification
DFN	debt-for-nature swap
EAP	Environment Action Program (World Bank)
FAO	United Nations Food and Agriculture Organisation
FMP	Forestry Master Plan (Regional Development Bank)
FSR	Forestry Sector Review (World Bank)
GDP	gross domestic product
GEF	Global Environment Facility

GIS	geographic information systems
ICRAF	International Centre for Research in Agroforestry (Kenya)
IDRC	International Development Research Center
IFAD	International Fund for Agricultural Development
IFS	integrated forest strategy
INCRA	National Colonization and Agrarian Reform Institute (Brazil)
IPF	Intergovernmental Panel on Forests
ITTA	International Tropical Timber Agreement
ITTO	International Tropical Timber Organization
IUCN	World Conservation Union/International Union for Conservation of Nature
IUFRO	International Union of Forestry Research Organizations
KEFRI	Kenya Forestry Research Institute
KENGO	Kenya Energy and Environment Nongovernmental Organizations
KREMU	Department of Resources, Surveys, and Remote Sensing (Kenya)
MABP	Man and Biosphere Programme (UNESCO)
NAP	national agriculture program
NBS	national biodiversity strategy
NCS	national conservation strategy
NEAP	national environment action plan
NEP	national environment program
NFAP	national forestry action plan
NFMP	national forestry master plan
nfp	national forest program
NFSR	national forest sector review
NGO	nongovernmental organization
NTFP	nontimber forest products
ODA	Overseas Development Agency (U.K.)
OECD	Organization for Economic Cooperation and Development
SAREC	Swedish Agency for Research Cooperation with Developing Countries
SFM	sustainable forest management

SHD	sustainable human development
TFAP	Tropical Forestry Action Programme (FAO and others)
UNCED	United Nations Conference on Environment and Development
UNDP	United Nations Development Programme
UNEP	United Nations Environment Programme
UNESCO	United Nations Education, Scientific, and Cultural Organisation
WCED	World Commission on Environment and Development
WRI	World Resources Institute
WTO	World Trade Organization
WTP	willingness to pay

CHAPTER ONE

Introduction

RALPH SCHMIDT, JOYCE K. BERRY, AND
JOHN C. GORDON

The United Nations Development Programme (UNDP) provided the idea, impetus, and funding for the preparation of
this book. The original perception was that there was a need
in the development community for a book on deforestation
that included information on both people and forests and focused on the creation of "country strategies" to reduce deforestation and fight poverty.

In order to test this perception, a questionnaire was given
to UNDP resident representatives (res reps) soliciting more
advice about the book's usefulness, form, and content.
Thirty-seven res reps from Africa, the Asia-Pacific region,
and the Latin American–Caribbean region responded. The
res reps confirmed the appropriateness of the publication
and also provided important perspectives on forest resources in their respective countries. When asked to rate on
a scale of 1 (very important) to 5 (not important) how important forests and forestry are as an area of activity for their office, 28 (76 percent) of res reps indicated either "very important" or "important." Only one respondent said that
forests were not important. Fifty-seven percent of the re-

spondents from Africa said that forests and forestry were very important, compared to 46 percent from Asia and the Pacific and 30 percent from Latin America and the Caribbean (table 1.1).

The second question asked the res reps to identify (again on a scale of 1–5) how important various forest resources or uses are to their smallholder farm families (table 1.2). According to respondents, most important to smallholder farm families are fuelwood and water. The res reps also said that in addition to fuelwood and water, the following resource uses are important: soil maintenance, construction, shelter, biodiversity, and food. Least important to farm families, according to the res reps, is aesthetic beauty of forests. When viewed by region, by far the most important use in Africa is fuelwood (86 percent said "very important"). In the Asia-Pacific region, res reps ranked both fuelwood (62 percent) and water (61 percent) as very important. Sixty percent of the res reps from Latin America and the Caribbean indicated that water is most important. Regional res reps had especially different perceptions about biodiversity. Sixty percent of Latin American and Caribbean respondents rated biodiversity "very important," as did 54 percent of Asia-Pacific respondents. Conversely, only 14 percent of African res reps said that biodiversity was very important to smallholder farm families.

Table 1.1 "Overall, How Important Are Forests and Forestry as an Area of Activity for Your Office?"

| | Very Important | | | Not Important | |
	1	2	3	4	5
Africa	57%	29%	7%	7%	0%
Asia and Pacific	46	39	15	0	0
Latin America and Caribbean	30	20	30	10	10
All res reps	46	30	16	5	3

Table 1.2 "How Important Do You Judge Each of the Following Forest Resources or Uses to Be to Your Smallholder Farm Families?"

	Most Important			Least Important		No Answer
	1	2	3	4	5	
Food						
Africa	43%	36%	14%	0%	7%	0%
A, P	15	31	15	23	8	8
LA, C*	20	10	40	20	10	0
All res reps	27	27	22	13	8	3
Water						
Africa	36	43	14	0	0	7
A, P	61	15	8	8	8	0
LA, C	60	0	40	0	0	0
All res reps	51	22	19	3	3	3
Shelter						
Africa	21	57	7	0	14	0
A, P	15	31	31	8	8	8
LA, C	20	30	30	10	0	10
All res reps	19	41	22	5	8	5
Energy/Fuelwood						
Africa	86	14	0	0	0	0
A, P	62	23	0	15	0	0
LA, C	40	40	10	0	0	10
All res reps	65	24	0	8	0	3
Fodder						
Africa	7	36	29	21	0	7
A, P	31	23	31	8	8	0
LA, C	0	20	20	30	10	20
All res reps	14	27	27	19	5	8
Soil Maintenance						
Africa	43	36	7	14	0	0
A, P	46	23	0	23	0	8
LA, C	40	10	40	0	10	0
All res reps	43	24	14	14	3	3

(*continued*)

Table 1.2 (*Continued*)

	Most Important			Least Important		No Answer
	1	2	3	4	5	
Medicine						
Africa	29	21	36	7	7	0
A, P	8	38	38	8	0	8
LA, C	0	50	40	0	10	0
All res reps	14	35	38	5	5	3
Construction						
Africa	36	29	21	7	7	0
A, P	31	46	23	0	0	0
LA, C	20	30	40	10	0	0
All res reps	30	35	27	5	3	0
Aesthetic Beauty						
Africa	0	0	21	21	43	14
A, P	15	23	23	23	8	8
LA, C	10	30	10	30	20	0
All res reps	8	16	19	24	24	8
Biodiversity						
Africa	14	26	14	26	7	7
A, P	54	15	8	15	0	8
LA, C	60	30	30	10	0	0
All res reps	41	16	16	19	3	5

*A, P = Africa and Pacific; LA, C = Latin America and Caribbean.

The concept of the book was welcomed, and many suggestions for improving its content were adopted from the survey. At the same time, we asked an international panel of distinguished and knowledgeable forest experts to comment on the book's concept and outline. They also supported and improved our starting point, and many became authors of chapters in the book. Every book should have a unitary ultimate use and goal. Ours is to help the development community break the negative spiral of poverty and deforestation.

Perhaps the best way to envision what might replace this negative spiral is to contemplate a positive vision for forest people on a local level:

> Imagine a rural community in a developing country: two thousand people living in a moist mid-elevation valley in any one of more than sixty countries.
>
> The community consists of farm families living on marginal agricultural land—land unsuited to intensive large-scale agriculture, land characterized by a mosaic of crops, forests, and other uses. One-half of the people of Africa, one-third of the people of Asia, and one-quarter of the people of Latin America live like this. Altogether, more than one and one-half billion people live on this kind of land.
>
> On the lower slopes and flatlands a variety of crops grown on family farms provide the main source of sustenance and income. Above these farms, the slopes become steeper and are covered by dense forest two thousand hectares in extent.
>
> The farming community knows that any attempt to crop in this forest would only be productive for a couple of years. The community's use agreement with the government for this two thousand–hectare forest is based on a simple management plan that is renewed periodically. The stream issuing from the forest flows year-round despite the four-month dry season. A series of small ponds produces fish for 100,000 meals each year.
>
> Half of the forest is of lesser slope and can be logged. One hundred people in the community work on logging in the dry season, when less work is required in the fields. Every year, one thousand trees are cut, yielding two thousand cubic meters of lumber. The lumber is sawed in a small mill owned and operated as a partnership by community members; the market value of the lumber is U.S. $180,000 annually. This includes some especially valuable veneer wood. A number of small furniture workshops use about one-third of the lumber; the market value

of what they produce is $120,000 per year. [All dollar figures are stated in U.S. dollars.]

Every household in the community gathers fuelwood from the forest. Only branches or trees not suitable for lumber are collected. This is a "free" good, costing only the labor involved, but it is vital for the health and well-being of the community. The value of this fuelwood based on nearby urban prices is equivalent to $80,000 per year.

The community hunts and traps birds and animals in the forest. Wild tubers and fruits provide 20,000 meals each year. The nearest doctor and source of medicine is a day's journey away, but several natural healers in the community who collect roots, leaves, and seeds for medicinal purposes are visited by community members every day. Palms and bamboo within and next to the forest are cut for thatch and poles.

Trees are found not only in the forests but throughout the farms. These farm trees provide fuelwood and fodder for animals, form live fences, hold soil and fix nitrogen between crop rows, produce fruit, and shade livestock and dwellings.

The community forest borders a larger national park on the far side of the ridge. A small lodge for visitors to this park collects $100,000 per year. The park and lodge provide a few jobs for the community, including one for a young guide who has been able to be educated. (Ralph Schmidt, personal communication)

From the standpoint of what is possible, technically or in human resource terms, no element of this positive picture is the least bit remote or unlikely. Yet reality is rarely, if ever, so idyllic. Social and economic pressures have worked against it. But these benefits are not imaginary. Various aspects of this scenario can be found in almost all rural communities in developing countries. They are commonplace survival mechanisms; for one and a half billion people they form part of the natural social safety net.

Forests provide water, food, protein, shelter, medicine, fuelwood, fodder, soil, lumber, and income from tourism. Some of these goods and benefits can be obtained, at least temporarily, from other sources. Many disappear altogether when forests decline. In the physical and social setting of most developing rural communities, the disappearance of the forest will greatly curtail the possibilities and options for sustainable human development. The human safety net disintegrates with the forest, poverty becomes more dire, and forest disintegration is accelerated.

We hope that this book helps the people in and around forests, and thus the forests themselves, by assisting a better channeling of development funds and energies.

CHAPTER TWO

Forests, Poverty, and This Book

JOHN C. GORDON, JOYCE K. BERRY, AND
RALPH SCHMIDT

Destruction and degradation of forests proceed throughout
much of the world despite vigorous efforts to combat them
(UNDP 1994a). New approaches are obviously needed, and
this book describes ways to create "integrated forest strategies." It is primarily intended for use by people combating
deforestation in countries where it is most severe and, of
those, particularly the ones that contain threatened moist
tropical forests. The book is mainly intended for policy makers, advisors, and forest decision makers, in both national
governments and international agencies.

The Need for Integrated Strategies

The Intergovernmental Panel on Forests provides a
framework for an international partnership to better understand and promote the sustainable management of forests. Its
creation signals a seriousness of purpose among the nations
of the world in the face of accelerating deforestation. Possible central mechanisms for achieving its purpose of understanding and combating deforestation are (1) the creation

and implementation of plans for the sustainable management of forests by the individual countries where deforestation is severe ("integrated forest strategies") and (2) a new consortium of industrial countries and multilateral agencies committed to providing major financial support for the strategies based on country-by-country discussions. In this approach, the creation of a forest strategy by a country experiencing high deforestation rates is the essential first step. The strategies will be country-specific in every respect but will serve international and global needs by combating deforestation to the benefit of all. An understanding that a better-integrated approach is needed can be gained by a glance at figure 2.1, which presents a summary of forest-related development efforts in one country. The overall picture, while portraying much laudable effort, is fragmented, convoluted, and obviously burdensome for the country being "helped."

A specific chain of logic underlies this approach and this book. Beginning with the fact that deforestation is proceeding rapidly in most developing countries containing moist tropical forests, and that this will cause increasingly severe economic, social, and environmental problems, the first task is to describe deforestation's root causes so that effective "cures" can be found. Many factors contribute to deforestation, and there are differences among countries and regions in the balance among deforestation's causes. For example, deforestation in relatively dry places (such as Haiti or Pakistan) often has fuelwood gathering and grazing as major proximal causes. In moist forest countries, shifting of cultivation and conversion to "permanent" agriculture, often preceded by exploitative logging, are more likely to be the apparent causes (figure 2.2). But the central theme everywhere in developing countries experiencing high rates of forest loss

Figure 2.1   The network of global aid and support available
to a country's national forest program

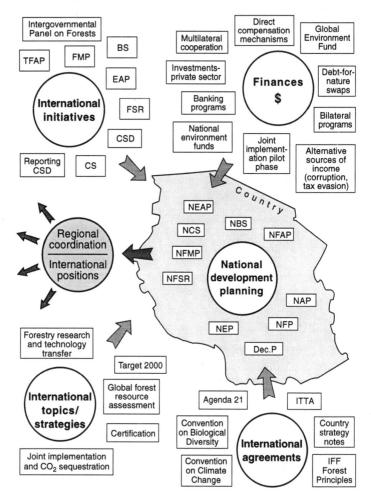

is the poverty of, and lack of options for, the people who live in and near forests or who migrate to them in search of a better life.

Most current efforts to reduce deforestation neither effectively focus on these people nor base their programs on thorough analysis of the specific goals and nonforest characteristics of the country being treated. Most do not address the broad societal influences from outside the forest sector that shape the relation between a country's forests and its people. Also, often in the name of the urgency of the problem, development assistance agencies and countries don't sufficiently coordinate their efforts with each other or with the countries they seek to assist.

Past efforts have been highly valuable, however, not only

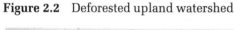

Figure 2.2 Deforested upland watershed

in beginning the enormous task of rolling back deforestation but principally because of the lessons they can teach. Foremost among these are (1) the realization that human resources, in the form of informed and empowered citizens, general and specifically technical education, and informed policy processes, form the basis for overcoming deforestation's more obvious drivers, and (2) the recognition of the need to revise the classical methods of valuing development efforts to better reflect the effects of resource depletion and environmental degradation or improvement. It is now clear that the integration of ideas from a variety of disciplines and sectors, together with a holistic view of landscapes (that includes all land uses and their interplay, not just forests), will be the hallmark of successful, sustainable forest development strategies. Integration across sectors of the economy, across geographic and political boundaries, and among culturally and economically different human populations is an enormous challenge but a necessary characteristic of realistic strategies (Gurderion, Hollings, and Light 1995).

It follows that efforts to foster sustainable forest development and combat deforestation can be most effective by assisting in the creation of strategies that focus on people ("human resources") that are led by the country to which they are to be applied and that are funded through carefully crafted and coordinated partnerships with a spectrum of development assistance agencies appropriate to the country and to the problems at hand.

Adaptive Ecosystem Management

Integrated strategies are tailored to specific country objectives and conditions using the principles of adaptive management of forest ecosystems. The essence of this approach is to learn by doing, in recognition of the complexity of forests and our relative ignorance of their detailed workings (Gurderion, Hollings, and Light 1995). The approach also implies that each strategy will be different because of differences in country goals, cultures, and economies and because of differences in the location, composition, and dynamics of forests. The overarching goal of this approach to management is, however, the same everywhere: the maintenance (and restoration, where necessary) of the health and functions of forests that are chosen to be retained or reestablished.

Gordon (1994) suggests that four broad themes define ecosystem management as a means to achieving the health and sustainability of forested lands, in addition to the classical scientific and technical bases of forest management. They are:

1. Manage with people in mind. People now affect, and are affected by, forests everywhere, through human impacts on the atmosphere, trade, and international relations. Thus even the remotest, least inhabited rain forest is subject to human influence that must be taken into account in its management. In fact, most extensive forests, particularly in the moist tropics, are inhabited by local or migratory people, and many are heavily populated. It is now clear that unless these local people are actively involved, management plans fail to achieve their objectives (see Chapter 4).
2. Manage where you are. Because of the need to manage large tracts of forest with few human or other resources, foresters have a tendency to use management techniques and silvicultural information very broadly. This has often led to the use of management approaches, harvesting systems, and ecological "facts" that are outside their useful domain, either geographically or with respect to specific objectives. For example, forest history is littered with "miracle trees" that,

having performed well in one or several places for one purpose, are widely distributed and used, only to be found to be economically and environmentally substandard and often to become weeds.

3. Manage with site-specific information. Forests are complex and ecological theory is weak. Thus, it is necessary to develop much of the specific information needed to manage a given forest in that forest itself. This has long been recognized in the local construction of volume tables and soil maps, for example, but it has become much clearer and more urgent with the widespread use of computer-based forest resource management and harvest scheduling models. The general knowledge base for tropical moist forests is particularly weak with respect to environmental protection functions and nonwood forest products, and this makes site-based planning imperative. The two major sources of local knowledge are, first, the people who have first-hand, often very long-term, knowledge of the forest, and second, adaptive processes in which every management activity is viewed from the perspective of knowledge acquisition and in which experimental designs and monitoring techniques are used to capture the lessons learned in day-to-day work.

4. Manage without externalities. Here, *externalities* are forest components that are ignored in planning and management. Forest management in the past has had, as its first activity, the choosing of the components that are of interest to the manager. Most often, this has been the forest's capacity to produce wood of particular kinds (sawlogs of certain species, fuelwood, mine timbers). Although this approach seemed efficient, in many places forest attributes of great value were diminished or lost before their worth was recognized. The ecosystem-adaptive approach requires that all known forest components be inventoried and that their fate be contemplated when making management

decisions. Of course, this is more expensive in the short term than focusing just on the attributes that are of immediate or commercial interest. Also, it is usually impossible to know each component, even in a sampling sense. But in the longer term, major value loss can be avoided, and, over time, a broad information base can be built. In the Pacific Northwest of the United States, for example, red alder, an early successional species, was systematically (and expensively) being eradicated as a weed in many places just before it became highly valuable as a timber tree and a provider of fixed nitrogen to forest soils. The situation is improving, according to a recent Intergovernmental Panel on Forests (IPF) report (IPF 1996 n.p.), which states that "although only a few countries are collecting information on the environmental functions of forests and plan non-wood benefits, those benefits are well recognized in many countries and recognition is growing."

The most important component of this approach to management—and the critical ingredient for developing an integrated forest strategy for a country—is vision. Developing a countrywide, broadly supported view of the future of the country's forests is the necessary first step toward the simultaneous reduction of deforestation and the achievement of sustainable forest development. Because this is essentially a political act, the methods used to develop the vision will vary with the cultural conditions, political processes, and institutions of individual countries; in every instance, the vision will not be developed solely by specialists but will reflect widely supportable ideas about the forests' role in the larger contexts of development and environmental protection. The same report (IPF 1996) urges that a high priority be given to capacity-building with a focus on the development of national strategies for the management of forests.

Creating an integrated strategy to pursue the vision requires that as much as possible be known about the interactions of the forest with other sectors of the economy and national life. It is especially important that all major beneficial aspects of the forest be recognized, along with the consequences of the conversion of forests to other uses. Most of all, it requires a whole (ecosystem) view of the forest and its contents and functions, particularly humans and their needs and impacts.

Forests and Poverty

The State of World Rural Poverty (IFAD 1992) presents the most comprehensive estimates we have found of this dismal situation. Of the population of 4 billion in developing countries (at that time—now about 4.8 billion), more than 2.5 billion were said to live in rural areas. Of these, approximately 1 billion were said to live below the poverty line: 633 million in Asia, 204 million in Africa, and 76 million in Latin America and the Caribbean.

Although the percentage of rural people living below the poverty line has decreased in several developing countries, the absolute number of the rural poor has been increasing. Sixty percent of these rural poor are women, 52 percent are smallholder farmers, 24 percent are landless, 6 percent are nomadic pastoralists, 7 percent are indigenous people, 4 percent are artisanal fisherfolk, and 6 percent are refugees.

We do not know precisely how many of this one billion rural poor live in or near forest lands, but we submit that the number is substantial. What many of us think of as "the environment" in developing countries is the survival base for these hundreds of millions. The IFAD report concludes that it is not a matter of chance that many (though not all) poor people live in areas of extreme environmental fragility (figure 2.3). Almost all such areas in developing countries are those where the presence of trees and forests is the only known means of sustainable utilization of the ecosystems.

The IFAD report then ties renewable resource management to development overall, stating that in more appropriate use of natural resources, involving local people, lies the answer to the challenge of development.

As juxtaposed in the compelling title of Nancy Peluso's book (1992b), rich forests and poor people are often found together, but when those rich forests are gone the people are even poorer. There is a common misconception that environment may be separated from human development, and this leads to a view that forestry issues and poverty issues

Figure 2.3   Slum dwellings in a forestry district

have little in common. This book argues that each set is central to the other.

In countries such as Haiti, Ethiopia, and Bangladesh, where forests are decimated over large and populated areas, the dire lack of forest products and benefits of all kinds is a major parameter and indicator of the impacts of poverty. Poor rural people who perceive these impacts know that overcoming lack of fuel, water, and medicine is among their highest priorities. Where ecosystems are not irreversibly degraded, the rural poor reverse extreme deforestation by producing their own tree crops.

Poverty is rightly viewed as one among many causes of deforestation (Patel-Weynand 1996). With no better alternatives, forest destruction and degradation become short-term solutions to many of the burdens imposed by poverty. It is equally true, however, that for poor people who live in and near forests and those who live in places where woodlands have been destroyed or degraded, silviculture can be effective in reducing poverty and ameliorating its effects on people and on the landscape by providing increased and sustained employment in the production and processing of timber and nontimber forest products. Beyond this, trees and forests can be important means of capital formation for those with little or no access to banks or to credit. In many places, trees, with their perennial habit, are sources of stable production through drought and other periods of climatic and market adversity. Unlike annual crops, wood harvests can often be delayed to coincide with need or better market conditions. Trees and forests, by providing shade, fodder, soil stabilization, and good watershed conditions, can improve the quality of life of those struggling to escape poverty and provide for many of the spiritual and aesthetic needs of local people (figure 2.4).

It is thus as important to view trees and forests as effective potential tools for poverty alleviation as it is to understand the connection between poverty and deforestation. This view of forests as development assets that are particularly relevant to the poor is sometimes seen as an antienvironmental stance. But most now agree that unless poverty is alleviated in and around forests, they will continue to be lost.

Figure 2.4   Reforestation project

It should also be clear that the sustainable use and reestablishment of forests is the closest tool (in both time and space) to use in fighting poverty among their inhabitants. Many opportunities exist to use specific forestry and agroforestry practices to enhance the economic and spiritual well-being of local people. The resources to do this on a large scale and the coordination required to avoid major errors and crossed purposes, however, will only arise from careful and integrated planning.

Conclusions

Thus the steps to use forests as a poverty-fighting tool are:

1. Recognize at the national and local levels that forests are a primary poverty-fighting asset and begin to view and describe them in that way;
2. Create an integrated national strategy for forests with this view at its heart;
3. Use the strategy to create partnerships that produce the resources for action and that reduce the current fragmentation of effort; and
4. Implement the strategy with broad participation and careful monitoring and feedback in an "adaptive management" approach.

Overview of the Chapters to Follow

In the following chapters, major themes underlying the creation of integrated forest strategies are examined and explained. Chapters 3 and 4 explore the lives and roles of people who live in and around moist tropical forests, who are often key and always important to successful strategy development. Only if their condition, motivations, and goals are understood and taken into account in the development of a strategy will there be a chance for its success. In the spirit of adaptive management, Chapter 5 examines what we have learned from past initiatives and explores the institutional settings needed to apply these lessons. The central one is that deforestation remains a severe and, in many places, an accelerating problem despite substantial efforts to reduce it. But frameworks such as the Tropical Forestry Action Programme and many country-based initiatives have brought a much better understanding of the ingredients of potentially successful forest strategies. Past efforts have also inculcated into the development assistance community a sensitivity to the need for better coordination among themselves and greater awareness that leadership must lie with the country affected rather than with development assistance agencies. Chapter 6 describes the kind and quantity of human resources necessary for the development and implementation of integrated forest strategies and stresses the need for these to be developed in the country and cultures of those whose vision underlies the strategies. Chapter 7 analyzes the economic causes of deforestation and suggests a new approach to the valuation of forests as a basis for strategy development. In particular, it examines the economic characteristics and motivations of those who live in and near forests and make their livings there. Chapter 8 brings together descriptions of the likely components of forest strategies and describes an approach to their practical preparation and implementation, and Chapter 9 provides answers to the key questions set forth below.

We have attempted to walk the fine line between useful specificity and ruinous overgeneralization. This is not intended to be a general text on forestry and economic development, and it is meant to be used with a wide variety of other materials (see the bibliography). It does present the rationale and methods for creating forest strategies. We offer this book in the hope that it will be instrumental in seeing that we reach the end of the twenty-first century with healthy and sustainable moist tropical forests and much less deforestation-associated poverty.

Key Questions

This book is thus structured to answer several key questions about the creation of country strategies:

1. Who lives in and around forests, and how do they interact with the forests, with each other, and with more remote actors, such as governments and businesses?
2. What institutional settings and approaches can be identified from past efforts to aid in the creation of new country strategies?
3. What human resources and what capacity-building activities are necessary for practical pursuit of new goals using new strategies?
4. What are the economic drivers of deforestation, and how can new methods, particularly of valuation, be used as a basis for strategy development?
5. What are the components of an effective country strategy, and how can they be integrated and implemented?
6. How do forest strategies connect with economic and social activities traditionally considered to lie in other sectors (agriculture, energy, manufacturing, international trade)?

CHAPTER THREE

Smallholder Farm Families in and near Forests

CHARLES BENBROOK

Converting a forest to a farm field is a time- and labor-intensive task. The job is made a little easier if logging or burning leaves the ground relatively free of logs, branches, and underbrush. Roads built for logging or mining also facilitate clearing. Still it is a long, difficult process, a chore no man or family takes on lightly.

Most freshly cleared forest soils are able to produce a few good harvests. For a while, families are well fed. But all too soon weeds gain a foothold and nutrient deficiencies or pest problems surface. Yields fall and other agronomic problems start to plague farmers, despite their best efforts.

Hungry families grow discouraged and think again about pushing on to find a better life. Often parents or parts of families stay but others move on. Almost always the road leads either back into urban centers or deeper into the forest and farther away from highly productive farming regions.

Forests and Agriculture

The FAO projects that the world's area in cropland will need to grow from 600 million hectares in 1990 to 720 million in 2010—an increase of 120 million. Of this, FAO projects, 85 million hectares will be converted in the countries that contain virtually all remaining tropical forests. Although it is not known what share of the 85 million hectares will come from forests, it will surely be a significant portion. If crop yield does not improve as projected, then substantially more land will need to be converted.

Table 3.1 shows that forests may be lost without comparable gains in cropland and that cropland can increase without comparable decreases in forestland. As people continue to move into upland areas, remaining forest resources often shrink fast, but this has little effect on national agricultural productive capacity.

This is clearly the case in Colombia, where only pockets of forested land remain in the mountain range between the

Table 3.1   Shifts in Cropland and Forestland Area Between 1979 and 1991

	Percentage Change— Cropland	Percentage Change— Forestland
Africa	5.0	−3.8
Gambia	14.5	−27.8
Ivory Coast	19.0	−25.8
Malawi	25.2	−22.8
Asia	1.3	−4.9
Thailand	25.5	−14.3
Bangladesh	2.1	−13.4
Philippines	2.5	−16.9
South America	12.7	−5.1
Brazil	23.1	−4.9
Paraguay	26.7	−31.6
Ecuador	9.4	−21.9
Former U.S.S.R.	−1.0	−22.0
World	1.8	−7.8

Source: Land Cover and Settlements (table 17.1), in *World Resources and Environment, 1994–95*. Washington, D.C.: World Resources Institute, 1995.

Cauca Valley and the Pacific Ocean. Large, mechanically managed acreage in the 925,000-hectare Cauca Valley has displaced both people and jobs, and so people have moved up the hillsides, where soils are fragile and few pockets of trees remain on very steep slopes. In some upland watersheds in Colombia and several other Central and South American countries, the loss of tree cover is triggering a host of water supply–related crises. Women are spending more time finding clean water in dry periods, which are lasting longer. Droughts are worsening.

During the rainy season there is little to hold the water in the uplands. It cascades down, carrying soil with it. Sedimentation clogs pipes, impairs fisheries, and fills rivers and reservoirs. In and around one small Indian village supported by one hundred hectares of cropland, 70 percent of the 1,200 millimeters of annual rainfall washed off the land, carrying with it 700 tons of silt per hectare to Sukhna Lake. The same situation is unfolding in countless other mountain regions, eroding lifestyles both in upland and lowland regions, uniting a country's people, who must confront and overcome this new and damaging form of environmental double jeopardy.

Upland watershed forest loss and erosion also force public utilities to struggle with a new, unanticipated set of problems they are ill-equipped to manage: dam storage capacity is lost prematurely, infectious diseases grow more prevalent, floods become more common and damaging, and periods of drought grow more severe and prolonged.

People and Agriculture

About 60 percent of the world's population lives in rural and forestland areas, and about 40 percent of the world's new inhabitants will settle in these areas. Forest loss is severe in many African and Asian countries because the supply of cropland is already low—less than two-tenths of a hectare per capita.

"World Agriculture: Towards 2010" (FAO 1995a) assesses the land-population balance in various regions. Sixty-five percent of the 1.2 billion people living in southern Asia, for example, depend on agriculture for their living. Eighty percent of the land suitable for agriculture is already in production. The forested area covers ninety million hectares, of which eighteen million hectares is judged to have crop production potential. This means that forestland is vulnerable, and most of it will degrade quickly once cleared despite the best efforts of farmers to hold hillside soils in place.

Pressures to Move On

Many pressures push people deeper into the forest, and most arise from outside forest margin areas: population growth, lack of opportunity in urban slums, new roads, policies encouraging new settlements, civil unrest, and natural disasters.

Other pressures to move on come from within rural and forest margin areas and stem from the need for fuelwood, new grazing areas, and unsustainable uses of cropland. In India, the annual demand for fuelwood is estimated at 240 million cubic meters, yet the sustainable production level is thought to be just 41 million cubic meters. Not surprisingly, the average time spent gathering fuel is rising in most areas. In several African countries and parts of southern Asia women now spend three to five hours per day gathering wood, and in some regions up to eight hours per day to meet the needs of a family of five (figure 3.1). Time spent seeking and packing wood limits time and energy for food production.

Unrestricted migration is a tradition in most of Africa. Roughly thirty-five million international migrants—half the world total—settled in sub-Saharan Africa in the 1980s. In northeastern India there are thirty million itinerant farmers who establish temporary food plots on forest margins. In 1985, about 70 percent of the upland population in the Philippines was composed of migrants from lowland areas. This shift was stimulated by several factors including the collapse of the sugar industry (a major source of jobs), the concentration of productive land in the hands of a few (5 percent of the farms controlling nearly 40 percent of the best land), and government resettlement programs.

Figure 3.1   Woman carrying fuelwood

Sustaining Soil and Farming System Productivity

Once cleared, forest soils can quickly lose their fertility and are subject to heavy erosion losses if the ground is not covered during rainy seasons. Farmers can combat erosion in two ways. First, cover crops, mulches, and other vegetation-management strategies can provide rainy-season cover as the forest did. Second, or simultaneously, appropriate land-forming conservation structures and practices (such as contour plowing and sod waterways) can slow down and channel rain.

Even on steep hillsides most once-forested land has a cover of productive soil, which is vulnerable when exposed. Annual erosion rates of more than 150 tons per acre per year are common—and remove about an inch of topsoil annually. Still, for a few years, even as they wash away, these soils can produce good crops. The productive potential of the land drops to one-third or less of that before clearing, however (figure 3.2).

Roads Often Pave the Way

The Cuiaba–Porto Velho highway (BR364) in Brazil, completed in 1968, made Rondonia state (a 243,000-km^2 federal territory mostly forested at the time) accessible for settlement and clearing. The government, in hopes of opening this new frontier, used the media in more populated areas of the nation to advertise the region's fertile soils and available government assistance.

Opening a new frontier was seen as an opportunity to lure people away from land in southern Brazil, where mechanized agriculture had reduced farm labor needs and where larger landholdings prevailed and smaller ones dwindled.

During the 1970s, in the southern state of Parana, the area in farms of less than 50 hectares fell by 890,000 hectares and the area in farms larger than 1,000 hectares grew by more than 1 million hectares.

In 1977 the National Colonization and Agrarian Reform Institute (INCRA) in Brazil established seven colonization projects providing hundred-acre lots and basic services to accommodate 28,000 families that hoped to develop small agricultural plots for cash and subsistence. An estimated 30,000 additional farm families could not be accommodated by the INCRA projects.

One of these colonization projects, located in Machadinho, Rondonia, was fraught with problems from the begin-

Figure 3.2   Deforestation through shifting cultivation leaves soil vulnerable to erosion

ning and has one of the highest incidences of malaria in the world. Machadinho is fairly distant from the BR364, which led the immigrants westward. The roads to it and within the project are often on hilltops, made of dirt and subject to erosion. The first plots of land were given to immigrants in late 1984 shortly before the rainy season began. The government established criteria for those eligible to get land and those who could not. These criteria proved adaptable to the whims of politicians, however, and often land was taken in exchange for services and goods by businesspeople who held it and sold it after its value had risen.

The tropical forest plots generally had nutrient-poor soils, and the immigrants found their land difficult to farm productively. The promises of physical and social assistance in the way of health, education, and agricultural extension were not fulfilled. Immigrants tried almost everything to live off their plots, yet generally returns were poor. Despite this, speculation continued. Often, land was held as collateral in supermarkets against immigrants' food bills, making their survival even more tenuous.

Several plotholders sent family members out to nearby *fazendas* (farms) to work and make money for their survival. Unfortunately, because Machadinho is quite a distance from the BR364, there was little access to better work and alternative moneymaking activities. In 1985 there were 295 farms in Machadinho, in 1986 there were 587, and in 1987 there were 835. In 1987 only 54 percent of all families that had settled in Machadinho in 1985 remained, because of a heavy turnover. Although land abandonment in frontier areas is common, this turnover is particularly high.

Life on the forest edge or in the tropical forest has not been the pretty picture that was painted for these immigrants.

Malaria has infected a large number of the newcomers. Promised government services have not been delivered. Food staples have not been sufficiently abundant to sustain the population. The highly acidic soils with high aluminum content have been difficult to farm, forcing people to move on multiple times in order to find better land or work. (Gold mines and logging camps have received these immigrants.) Often people anticipate using a plot for three years and then having to move on, fueling a pattern of forest clearing and impoverishment.

Places to Focus Effort

Export Crops on Fertile Lands

In many countries some of the richest farmland available is controlled by a few families or corporations and is used to produce export or nonfood crops using large-scale, capital-intensive production methods, often coupled with tasks that are very labor-intensive, such as harvesting cotton or sugarcane. Such farming operations typically create hundreds, even thousands of jobs, but they are almost always very short-lived and rarely provide people with a steady source of income. Nor do such farms contribute much to meeting national food needs. The land controlled by such operations, if used differently, could contribute greatly to food security and employment for the rural poor. But to achieve this shift, some government action generally is needed to change the opportunities open to landowners.

New policies or strong prices for certain food crops might convince farm owners and managers to change what they produce. Other concrete actions by governments, nongovernmental organizations (NGOs), or the development community might play a role in encouraging a shift away from capital- and energy-intensive methods and toward more diversified, high-value crops and enterprises that create many more year-round jobs. The end result can be dramatic. A thousand-hectare mechanized cotton, sugarcane, or beef farm might support perhaps two dozen permanent employees and an occasional workforce of several hundred. But if the best land were instead devoted to a combination of fruits and vegetables and small-scale but intensive livestock and fishery production, the same farm could easily support a hundred families year-round and twice that many in associated harvesting, packing, processing, marketing, and food manufacturing enterprises.

But such transformations rarely happen without some changes in external factors—a new road or port opening up new markets or another change in infrastructure that makes it possible to grow crops that previously were not profitable. By targeting support for infrastructure, market development, and education to such areas, governments can increase the number of landowners choosing to accept the risks associated with change.

Government and the development community can accelerate shifts in land use by creating new market demand for a different mix of crops and commodities. Such entities should strive to diversify a region's economic base and create jobs by offering assistance to people and cooperatives interested in starting manufacturing, processing, and marketing enterprises. Areas moving through such transitions typically become magnets for people on the move. Governments rarely have to do much to encourage people to move where there appear to be real opportunities.

Stabilizing Forest Margin Areas

Slash-and-burn techniques, both exploiting and expanding forest margin areas, have been used to shift cultivation
for centuries. Such patterns of land use can be sustainable
and do minimal damage to the forest as long as population
pressures are modest and forested areas vast. But this is no
longer the case in any developing country with significant
closed forest.

Anything that can be done to increase and sustain the productivity of forest margin areas and expand economic activity will give people a reason to stay. Often extra effort and investment are needed just to keep such areas from rapidly
degrading.

Developments in Zaire point to the complex interaction
of demographics, culture, and economics in shaping—or de-
laying—the transition to sustainable farming systems. Zaire
contains about 48 percent of all forests in central Africa. The
Ituri Forest, a closed forest in Zaire's central basin, is home
to the Mbuti aboriginal hunters and gatherers, the indigenous Bira farmers who have inhabited the forest for four to
five hundred years, and the Nande, inhabitants of the forest
margins who immigrated within the last five to twenty years
from population centers north and east of the forest.

Located in the North Kivu region, the forest is managed
through concessions, many of which are held by Bira using
their culture's traditional system of ownership. The Ituri
Forest tenure systems are somewhat informal and unstruc-
tured, making this region inviting to immigrants. Land is
abundant and easily acquired from the Bira, unlike the previous homes of the Nande, where high population density
and unequal distribution of landholdings made survival dif-

ficult or impossible. Once newcomers are settled and sur-
viving, their extended families often join them, pushing further into the forest and occupying traditional Mbuti and Bira
land.

The agricultural practices of the indigenous Bira farmers
are generally not adopted by the Nande. Immigrants generally try to increase production by augmenting plot size and
intensifying production, depleting both forest stands and
soils.

The Bira farm small plots (on average .9 hectares) of manioc and rice, and traditionally they allow land to go unculti-
vated for five to six years before putting it back into production. They cultivate plantains in the primary forest soils. By
comparison, the Nande immigrants, accustomed to the or-
ganically rich and resilient volcanic soils of the northeastern
savanna, utilize the land more intensively and allow plots to
stand fallow for a maximum of three years. The Nande also
grow cash crops for whose production forests are perma-
nently converted to agricultural fields. Although anecdotal
information reveals that productivity on more intensively
used lands has fallen, the Bira see that the Nande consis-
tently produce good food and cash crops on their farms.
Some Bira believe that their agricultural practices are back-
wards and wish to follow the Nande ways by farming more
of their land more often.

The Bira plant their crop and wait for its harvest. The
Nande plant one crop and then busily add second or third
crops to their landholdings. The Nande consider the Bira to
be lazy and unmotivated, and it seems that young Bira have
little respect for elder Bira and their traditional agricultural
practices. Young Bira often abandon agriculture to mine gold
at camps along rivers throughout the Ituri Forest. In this way
the indigenous farming traditions, which have had little im-

pact on the forest during the last four hundred years, are being lost.

Many Nande see coffee production as their path to success in the Ituri Forest region. Coffee is increasingly becoming a cornerstone of economic welfare, generating disparity in the standards of living of indigenous farmers and immigrants. Immigrants live in painted houses with tin roofs, own better farm tools, and have more livestock and increasingly more political power than the indigenous Bira farmers. As the Nande are able to improve their well-being with cash from coffee (as well as manioc and cattle), the Bira become more interested in clearing additional land for cash crops.

The Nande people came from a savanna environment very different from the Ituri Forest. They knew little about the forest resources and products on which the Mbuti have always depended and which the Bira know and utilize. Immigrant farmers have little interest in learning about the forest and believe that it is really of no use to them. As Nande immigrants told an American researcher, Richard Peterson, during a study, "The only benefit of the forest is land to farm and food that is grown to sell and eat." "We do not go to the forest. I do not get one thing from the forest. The profit of the forest is farming." "There is nothing in the forest that one can live on. The only food of the forest is that which I put into it." "I don't know about the things of the forest. Things like forest fruits, we are not accustomed to." "Getting things from the forest is the concern of the local people only."

Change in farming practice is due not only to ethnic differences but also to increased population density. Since 1975 thousands have come into the Ituri Forest to live. Many have stayed and more are bound to follow, especially now, as health concerns spread in urban centers.

Creating Opportunity in the Forest's Shadow

The policy implications are certain: forest clearing is not going to solve the food production problems and rural poverty of developing regions. Most of this will have to be done through increasing the productivity of other lands. If something is not done quickly and decisively, in several countries, the first stages of development will soon begin reaching into the last remaining closed forest regions; within a generation new pressures will follow, and there will be very little closed forest left to lose and almost nothing gained as a result.

Agricultural enterprises must become more rewarding, reliable, and sustainable for rural peoples if they are to have hope and options. Farmers worldwide understand the importance of controlling erosion. They know that soil nutrients have to be replenished and that pests have to be managed if there is to be much left to harvest. But too often they are unable to figure out how to accomplish these essential goals while still feeding their families. New tools, technologies, and inputs combined with knowledge and guided by agroecological principles can set the stage for rapid progress. But knowledge must come first and in the long run is a far more valuable asset than any input or implement.

The knowledge, skills, and inputs needed to shape agroecological systems differ markedly from those considered essential in past agricultural development projects, most of which were designed on the premise that the surest path to increased productivity was through a package of inputs and technologies. Such engineering-based, input-driven systems are prone to biological problems even under the best of circumstances; they are doomed to failure in forest margin and degraded regions. The best way to develop, adapt, and then transfer such systems to farmers is to expand reliance on farmer-driven participatory research, demonstration, and training .

Summary

Conversion of tropical moist forests to agricultural use, usually shifting cultivation, is a major land-use dynamic today. The United Nations Food and Agriculture Organisation projects that 120 million hectares of new cropland will be converted between 1990 and 2010, though how much of this will be from forests has not been estimated. Converting a tropical forest for a temporary crop is a difficult and unhealthy task that will produce only a meager reward; few will do it if they have alternatives. Major agricultural policy issues— such as incentives for intensive cash cropping or road building, or land ownership laws and practices—have a profound impact on pressures that push people deeper into the forest.

There appear to be a number of policy actions that can improve agricultural production, increase food security and income for the rural poor, and reduce unsustainable conversion of forests to temporary agricultural production. These include labor-intensive use of high-potential lands for production of food staples. Very little attention, in terms of either research or extension, has been devoted to shifting cultivation. Improvement and stabilization of this kind of agriculture could also have a major impact on food security, poverty reduction, and sustaining of forests. Work with smallholder farmers will always occur on a local scale. The patterns of agricultural practice, gender relations, and culture are always intricate, diverse, and complex. Effective action requires more than an understanding of this; it will occur when the local culture is empowered and enabled to harness its own efforts for sustainable development.

CHAPTER FOUR

The Role of Forests in Sustaining Smallholders

NANCY PELUSO

A forest is a social and political place. Resource users, including smallholders, government management agencies, conservation groups, and timber concessions, have diverse and often conflicting interests in how forests are managed. Complicating their management is the fact that each of these groups is itself heterogeneous, comprising people with different perceptions of and interests in the forest. These people are also embedded within a variety of social relations that influence the ways they will interact with—manage and use—the forest. Thus, people's relations with others are as important to understanding their use of the forest as are their direct forest management activities. Forests can become sites of contestation where social and political struggles are played out. These struggles often derive from conflict over access to the forest and the formal and informal means by which people gain that access.

Resource access is a multidimensional concept that lies at the heart of many resource-related problems. *Access,* as used here, means more than simply "tenure" or "property rights." Besides geographical constraints on physical access to a resource—the capacity to reach it on foot, by boat, by road, or by rail—social mechanisms also channel people's access through resource tenure arrangements, market mechanisms, or labor relations. Each of these mechanisms is defined below.

Resource tenure arrangements refers to the formal and informal means by which people make claims to resources. The word *tenure* includes both formal property rights, recognized in formal-legal or customary laws, to a particular resource or set of resources, and informal or unrecognized claims to the same set of resources. *Formal* implies the legal recognition of the rights by a state or other authority and the willingness of that authority to enforce a particular set of claims. Informal rights are claims that are sometimes recognized in practice, that is, by the simple act of taking or using a resource, even though local or external authorities do not formally recognize its use. Thus, although in common parlance *property* is often used to denote the relation of a person to a thing, in fact it defines the relations between a person and all other persons in relation to that thing; property is a social relation by which a person, a group, or a corporate entity can exclude others from the use of a particular thing (Macpherson 1983).

Market mechanisms affecting resource access also have formal and informal components. Formal mechanisms include the variety of permits for production, extraction, or transport of resources, often those that are handled by a government or other formal authority. Also included here would be tariffs and pricing policies, as these might constrain a user's ability to enter the market. Informal mechanisms include underground markets—whether gray or black—and their provision of marketing channels that serve as alternatives to those approved by the state.

People may also gain access through a variety of labor arrangements. Thus, even though a group may have no access to a resource through the tenure system, and may not have the capital to engage in commercial transactions giving them possession of a resource, they may gain access by entering into a working relationship with the owner. Through their labor, they are then able to acquire some of the benefits of exploiting the resource, in the form of either cash payment for their labor or a percentage of the harvested material in kind.

The remainder of this chapter will focus on resource tenure arrangements as a window to understanding the variety of ways in which forests sustain smallholders. This requires examining the differences in the means people can deploy in order to gain access to resources. It also requires understanding differences as they are expressed in local social relations and how these are connected to outside influences such as national laws and policies, markets, and social or environmental changes (Blaikie 1985). The interactions of any of these levels of management—state with local, local with international, or international with state and local systems—can lead to unpredicted and unpredictable outcomes for both the resource base and various resource users. Questions of access and how rights are perceived, recognized, ignored, appropriated, established in law, and contested are the major problems facing forest managers today.

A Political Ecology Approach

A political ecology approach to analyzing the ways forests sustain people's livelihoods—and the ways these livelihoods are threatened—can help the analyst identify critical processes, structures, and relationships affecting forest use (Blaikie 1985; Blaikie and Brookfield 1987; Schmink and Wood 1987, 1992; Neumann 1992; Peluso 1992c). Political ecology is an approach to research or evaluation that examines the social origins of environmental problems by studying multiple versions of local, regional, and national history; analyzing local social relations, that is, the interactions of resource users with the environment and with each other; and contextualizing local situations within larger political-economic, geographic, and environmental settings (figure 4.1).

In order to understand the multiple influences on a resource user's behavior, a political ecologist would start by identifying the user as well as the particular management practices or strategies, institutions for access and control, and sources of conflict and cooperation that are present.

Figure 4.1   Political ecology approach

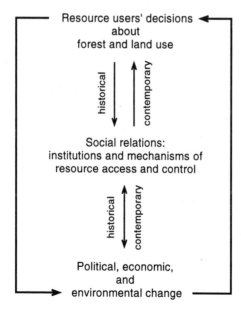

Then she would examine relevant state and market systems or aspects of forest management that have been superimposed on local or customary systems for the production of timber or other forest products. Finally, she would examine international influences on local forest management or use and the reverberations of local activities "outward" to the national and international levels. In addition to influences deriving from each of the various levels of analysis, unanticipated outcomes often result from the interactions between multiple resource management strategies. Local, state, and international systems of management can be thought of as a set of cumulative impacts on the ways individual people actually use forests.

Layers of Resource Tenure

Learning How Local People Claim the Forest

There is no such thing as a local management system that is divorced from state, market, and other outside influences. Some local systems may be only minimally influenced by external forces, however. For example, prior to the advent of colonial rule, people living in parts of the forest geographically remote from political centers had a great deal of autonomy in their everyday actions, largely because geographical barriers prevented effective supervision by kings, princes, or feudal lords (Reid 1988).

In spite of their distance from political centers, many so-called remote forest communities have a long history of contact with international markets. Traders from Africa, China, India, and later Europe sought products available in deep forests as long ago as two millennia (Hall 1985; Wolters 1967). The demands of external markets could change the nature of people's interactions with their forests periodically, causing them to leave their usual agricultural or subsistence forest-product-gathering pursuits to collect specific products with no local uses to trade for manufactured items or other exotic goods produced elsewhere.

Forest managers with the most sophisticated formal training need to understand how and why local people make claims to a forest and its products. This means not only learning what species people use, what kinds of products they develop, and whether they are used for commercial or subsistence purposes but also what sociological differences constrain or facilitate the activities of forest users. For example, variations in use are likely to be observed among different gender groups, socioeconomic classes, and age or ethnic groups. Poor people may be more dependent on access to certain labor-consuming forest products that better-off villagers are not willing to harvest, but when such products acquire a high commercial value, wealthier villagers may try to monopolize harvesting rights. Men may hold rights in land, whereas women hold rights in trees; in some cultures women are not allowed to plant trees. All of these sociocultural factors influence the effectiveness of an intervention in forest management and often change the forms that the legal terms of access take.

Given the inevitable differences within any particular forest-based community, a critical function of any program for sustainable forestry is the identification of locally important forms of social differentiation. As mentioned above, some key sociological categories can be examined systematically, including gender, age, income group, status (caste, leadership roles), ethnicity, and religion. The task of the forest manager is to discover first how people in these various groups use forests and trees and the nature of the relations that develop between them and people in other local and external groups. The next step is to learn the origins of these preferences and how the ultimate use of the forest is overtly or covertly negotiated between user groups.

The sociological differences between forest users derive from what we might think of as "social scarcities." That is, even where resources are not physically scarce, social barriers prevent some people from gaining or maintaining access to them. When people have little control over resources, they do not take care of them. Scarce resources are even more constraining. The most obvious social scarcities come from inequities in access between people in different economic classes or in different status groups such as castes or gender groups.

Trees as Weapons in the Gender War of Rural Gambia

In Kerewan, a Mandinka village in the Gambia, a struggle of land claims has been waged between men and women, both of whom plant trees and vegetables in order to make strategic gains. Resource tenure in Kerewan has always been dependent on gender, but recent ecological and development policies have interacted to fold women into a national cash-crop economy so that they have taken on more responsibility within the household and gained greater access to financial resources. This increase has come at the expense of men's control of cash within the household and consequently has encouraged gender-based social rifts to rise to the surface. These rifts are embodied in the conflict over the planting of trees in vegetable gardens and is characteristic of the changing political ecology of Kerewan.

Whereas the power relationship between women and men is dynamic, gender-based control over natural resources showed certain "traditional" characteristics twenty years ago. Traditionally, women tended lowland, swamp-based crops such as rice on land they or their mothers cleared. During the dry season, their lands provided the only significant production in the Mandinka labor calendar. Throughout the year, most of the product of these lands was used within the home. Men, on the other hand, were the tenders of the upland arable lands where groundnuts, millet, and maize were grown for sale in the market. Theirs was—according to the Mandinka tradition—the sole source of cash for the household.

In the early 1970s this dynamic started to change as women gained more access to cash resources, and the current conflict over land claims began. The change has roots in the evolving ecological conditions of a long-term drought and in the shifting priorities of international development programs toward women in development. During the years of drought, women adopted faster-growing rice varieties in accordance with the shortened rainy season. The two months that these varieties save freed women to work more on irrigated dry-season crops. International development agencies focused on increased dry-season production from women's vegetable gardens as a way to increase both production of vegetables and women's access to the cash economy.

This shift in local ecology has embedded within it a shift in local politics. It has led to a struggle of claims to control over the land on which these gardens are grown. As the importance of cash crop vegetables has increased, men have begun to plant trees as a way to reestablish their control over the cash economy.

According to Mandinka custom, trees belong to the planter, not the landowner. This tradition leads to conflict between male landowners and female gardeners. Women plant the dry-season vegetable gardens on land owned by senior males. In order to secure user rights, women make a one-time cash payment to the landowner, but since owners are not required to keep official records, they are able to regularly extract this payment as an informal kind of rent. These landlords restrict the women from planting mango, orange, or other woody species in the gardens because they imply long-term ownership. Instead, they allow shorter-lived papaya and banana trees to be planted among the vegetable species so as to maintain the strength of their claims to the land and the ability to control how that land is used.

For this reason, trees became a threat to women's tenure over the gardens they tended and therefore a threat to their link with the cash economy. The following example of one development project's support for increased women's dry-season activities is characteristic of the general conflict that has arisen over the planting of trees. The organization developed a project to expand women's gardens onto land owned by a local male landowner. The owner, although supportive of the general idea of the project, refused to allow the women to be direct cosigners of the project agreement on the ground that this would undermine his land rights. In response to his conditions' not being met, he blocked all expansion construction efforts and threatened to evict the growers from his land. The women not only refused to leave the garden but also developed a plan to tear down the old garden fence in preparation for the new garden. This conflict provoked the police to intervene and invoke a cooling-off period of several months.

During this time, the women lobbied national women's organizations and the development organization to intervene on their behalf. When the court ruled in favor of the women on all claims except those regarding woody fruit tree species they had planted, they were given access rights but ordered to remove the trees. Immediately following their removal, the owner planted several dozen of the same species in the expectation that the irrigation water the women would bring to the garden for their vegetables would also support his trees. These new trees died, and he claimed that the garden had been deliberately abandoned by the women in retribution for his previous heavy-handedness in the whole controversy. The man did not attempt to replant the trees after the garden's abandonment and was satisfied that he had reestablished undisputed control over his land. His actions created a precedent for other landholders to use female labor to water their own private fruit tree gardens.

By the mid-1980s, the relative economic benefits of fruit trees had shifted in favor of vegetables, however, and trees had—in fact—become a threat to women. As a result, vegetable growers began cutting back and chopping down trees in order to provide sunlight for better vegetable growth. Simultaneously, landowners saw a growing incentive to plant fruit trees since developers were eager to support tree planting under any conditions as a means of land stabilization. Perhaps unbeknownst to developers, this encouragement to plant trees favored the men, who could "capture" female labor to water, manure, and guard the landowners' trees while they were tending their own vegetables. These latest efforts by landowners have been strongly resisted by the growers, who regularly burn and remove the fruit trees in the gardens.

In Kerewan, trees are not apolitical. They have become a means by which different parties gain both symbolic and material control over resources. Gardens have become the battlefields and trees the weapons in the conflict over gendered political relations.

It is often said that poor people tend to be more depen-

dent on the forest than better-off people because the poor
have access to fewer privately held resources, especially to
the most valuable of resources, such as land. According to
this argument, forests, often seen by local people as their
common property rather than as the open-access resources
Hardin (1968) discusses, become refuges for the poor, who
may even work out common property arrangements with
each other as a way of managing the forest and allocating access to its products.

Trees and other forest products can serve as contingen-
cies for vulnerable households. Whether they are found and
protected or planted, trees help families through times of the
year that are invariably characterized by subsistence shortages—the hungry season that usually occurs in the months
or weeks preceding the first or only harvest of staple crops
during the year. Trees can also help vulnerable households
make it through times of family economic crisis and stress
such as illnesses, funerals, and marriages. Trees are also alternative productive investments, replacing other kinds of
"hard" investments such as cattle or gold jewelry. Trees may
be more desirable investments because they do not require
much advance capital to acquire. In moist forests they increase rapidly in value, and where the product is valued
more than the trees themselves, they may produce products
with high divisibility (Chambers and Leach 1989).

Multiple Influences on Tree Tenure:
Durian in West Kalimantan

Selako Dayaks of Bagak Sahwa, a village in West Kalimantan, Indonesia, have long planted durian trees (*Duriozi-
bethanus*) as part of their landscapes. Two to three generations ago, when shifting of cultivation was physically and

politically feasible, these Dayaks "shifted" only short dis-
tances from recent settlement sites. When they moved, they
took as many building materials from the sites as possible,
leaving a gap in the wooded areas that usually surrounded
their houses. They filled in these gaps by planting durian and
other types of fruit. The durian had special meaning for local people because they represented both the ancestors who
had planted them and significant historical events that took
place near them. Trees were actually named after both ancestors and events and sometimes had their names changed
during the course of their 100–150-year life spans.

Until the turn of the twentieth century, the Bagak villagers' landscape was dominated by swidden fields and fallows with patches of managed fruit forests. During those
times, rice self-sufficiency was crucial to local people's survival because markets were few and people needed large
swiddens to provide their family's food for the year.

Over the past century, and particularly since the late
1960s, villagers have transformed their local landscape in response to changes in both market access and government
zoning policies implemented since the 1930s. Today economically valuable trees—fruit and rubber—dominate the
landscape. As one woman said, "Rubber is our daily rice."
Fruit brings windfall profits in good production years. The
relations between land, trees, and the systems of tenure
around them have been revolutionized by changes in the region's political ecology, whose effects on social organization
and landscape in Bagak have included

- the sedentarization of villagers and formalization of
  village borders by the gazetting of a nature reserve, the
  establishment of a Catholic mission school, and the establishment of a rubber plantation and three resettlement areas along those borders;

- the hardening and subsequent paving of a tertiary road, increasing village access to urban areas, markets, and employment;
- the fluctuating enforcement of the nature reserve's borders and with it the fluctuating access of local people to the land and resources within those borders;
- the introduction of rubber by colonial planters and its appropriation by local people—a purely cash crop grown in smallholder plots; and
- the booming market for fruit, brought on by the improvement and expansion of transport facilities.

As land for swidden plots grew scarce and markets became more accessible, people chose to grow and sell rubber to buy or trade for rice. They started planting rubber in swidden fallows, shifting the landscape's composition. Rubber and fruit trees eventually came to dominate the hillside landscape, encroaching on swidden fields and less intensively managed fallows. This land-use revolution entailed an expansion of some land use categories at the expense of others and a change in the types of places certain trees are planted. Durian are now planted not only in former living sites but also in swidden fallows, invading rubber gardens (which were themselves planted in swidden fallows). By 1990, 85 percent of the villagers owned productive rubber trees. In addition, 97 percent of sample households planted durian trees in their swidden fallows; at least 41 percent planted durian in or next to their rubber gardens.

The longevity of the trees has changed the nature of property rights in land. These processes also changed the ways in which some people value the mix of resources. Villagers now manage a range of forest types, each with different origins, species compositions, uses, and combinations of property relations. The borders between these land-use types are blurred, and uses overlap. Moreover, these forest management categories are neither understood nor recognized by state forest managers in any official capacity. They are not mapped, for example, as "social forests."

People now talk about the trees as investments for their children's future. Whereas formerly fruit trees represented kinship and village history, today a new meaning has been overlain on them: they are commodities that bring significant profits in good production years. Whereas formerly the tree planter's descendants would share the harvest in common, with specific rules of access applying to each succeeding generation of inheritors, today many parents plant plots of land in the valuable trees and pass them on to specific children. Thus landscape change and political economy have altered the meaning and value of and property rights in these important trees.

When certain products in a forest increase in value, however, the social relations affecting other parts of poor people's lives may change the way they interact with that forest. Though poor people may retain their physical access to a forest, they may become impelled to sell or deliver the valuable products to villagers to whom they owe social debts—their patrons. Failure to do so may result in loss of access to work opportunities outside the forest. In such cases, the stresses on the forest resource are not caused by poverty—even though poor people may be engaged in extraction or production—but rather they are caused by processes of wealth accumulation engaged in by the more affluent villages.

Tree Planting Divides Large and Small Farmers

Under certain circumstances competing claims to ownership, both formal and informal, can result in the forest's becoming a site in which resource access is contested. A tree-planting project in Pakistan, for example, effectively gave wealthy farmers the means by which they could enclose commonly held grazing lands and exclude poorer farmers from needed resources. By inadvertently removing a channel of access to resources for poor farmers, the project set up a political dynamic whereby the resource would have little value for the majority of smaller farmers, who now had incentives to ensure the forest's failure.

According to the project's plans, trees were to be planted on government, private, and Shamlat community land. The Shamlat forests were intended to provide needed fuelwood to small farmers, and the seedlings were to be planted by those expected to receive the forest's benefits. Government land was planted to provide a model of the benefits of fast-growing forests in an effort to encourage landowners to plant trees on their private lands.

The project was successful primarily in the privately planted areas on both private property and Shamlat land. The only farmers willing to invest in planting on the Shamlat land, however, were the larger farmers, whereas smaller farmers—the ones intended to benefit most from the program—refused. The reason for refusal was based on the contested nature of ownership in the Shamlat areas. Although legally these tracts are commonly owned, in practice they are divided among individuals. Likewise, the benefits from the Shamlat lands accrue according to these subdivisions. Project designers assumed that these lands were communally owned, but over generations a small number of families had informally gained exclusive use rights to certain areas proportionate to the cultivable land already within the family. Thus, historically small landowners and landless farmers had little or no access to the supposedly communal land on which the trees were planted. Their rights to this land became limited to lesser but important secondary uses such as grazing livestock. In this way, an informal process of privatization had occurred, and the larger farmers hoped to get the Shamlat land planted at no cost to themselves and with no obligation to the government or the project's funders. They hoped to enjoy shared costs and individual benefits.

Not only would small farmers plant trees on land effectively owned by larger farmers, but the benefits of this planting would accrue only to those with exclusive control over the plots. In addition, the project also would have removed small farmers' rights to fodder and grass from the common land. In such a situation, if tree planting occurs on what is de jure common land but de facto private property, poor farmers stand to lose important grazing rights because planting will solidify the larger farmers' control over their individual plots. With the conversion of common grazing land to forests, the project—and indeed the forest itself—was placed in a precarious position where the majority of local farmers had compelling incentives to ensure its failure.

Forest Management Systems as Political Tools

The state of Para in the Brazilian Amazon provides a good example of how changing social and political relations determine access to a resource and consequently the way in which the resource is used. In this region of the Amazonian forests, the Kayapo Indians have gained control over their traditional lands through carefully calculated political investments. They have used their traditional forest management practices to gain the support of the international environmental movement. With this political support, they have been able to exert greater control over their traditional land claims than they had been able to using traditional weapons such as spears and integration into the greater Brazilian society.

In 1978 a government decree defined the Gorotire reserve, land intended to be ceded to the Kayapo Indians, who claimed the general area as their traditional lands. Within a year, however, surveying efforts bogged down. Concurrently, a land boom and road construction in the Amazon enticed to the area scores of cattle ranchers who began to encroach on what had been defined as Kayapo land. Consequently, the Kayapo took enforcement of the borders into their own hands by patrolling the perimeter and attacking persistent encroachers. Although this traditional form of establishing ownership over land killed some ranching families, it did not slow the influx of new ranchers, who were increasingly joined by miners and loggers in their search for "unowned" land. Where force failed, political investments have, to a certain degree, succeeded in establishing Kayapo claims to the forest at the expense of ranchers, miners, and loggers.

Conflict over gold mines in the reserve brought to a head the issue of competing claims to resources and provided a means by which the Kayapo established rights to their land over not only miners but ranchers and loggers as well. Even though the Brazilian government had set aside this land for the Kayapo, it simultaneously developed policies to colonize the Amazon by building roads, subsidizing migration to the Amazon from other areas of Brazil. These subsidies were intended, in part, to make the Amazon an economically productive region providing minerals, logs, and beef. The official declaration of the reserve was thus undermined by the tremendous incentives and opportunities provided for miners, ranchers, and loggers to encroach on the reserve. Government policy toward implementing protection for the reserve contradicted the overall policy of encouraging use of the Amazon's resources. The land was claimed by different parties that drew on different policies for their legitimacy. The Kayapo drew on the official 1978 decree while the encroachers claimed that failure to exploit the resources on reserve land was a block to economic development.

In response to miners staking out claims on the reserve, the Kayapo invested in a political resource with a less ambiguous agenda than the Brazilian government's. In the mid-1980s they tapped the growing environmental movement in the United States and Europe as a way to influence the Brazilian government to favor the Kayapo claims. In the early 1980s, Western anthropologists began to understand the traditional Kayapo ways of managing the forest as uniquely adapted to the physical environment of the Amazon and to see that these principles might be a model for sustainable development. This burgeoning interest led to the invitation of Kayapo leaders to conferences such as "Wise Management of Tropical Forests," sponsored in 1988 by Florida International University in Miami. Furthermore, these leaders had the opportunity to meet with officials from the World Bank, the U.S. State Department, the U.S. Treasury Department, and Congress. In these meetings the Kayapo expressed concern about proposals to build a series of dams along one of the rivers flowing through their territory. Kayapo access to high-level organizations funding Brazil's economic development in the Amazon had major consequences that the traditional forms of resistance could not attain. Most immediately, it prompted the Brazilian government to charge the Kayapo leaders with foreign sedition, but more important, it created a political alliance between the Kayapo and the environmental movement, with its growing ability to influence the world, which provided much of the money to develop the Amazon. Ultimately, this alliance led to the consolidation of

Kayapo land claims at the expense of the miners, ranchers, and loggers.

The security of Kayapo rights to land was the direct result of their investment in political groups outside the country with interests that coincided with their own. Environmental groups were interested in the preservation of the rain forest and saw the Kayapo management system as a model for the Amazon's development. The Kayapo used this interest not simply as a way to preserve the rain forest but, more important, as a way to establish their rights to their traditional lands.

Forested land is rarely unmanaged by humans. Moreover, multiple management systems often exist in a dynamic relationship, involving differing social groups. Forest tracts are contested zones where the type and degree of stress placed on resources is determined by the social and political relations not only among the groups living in the forest but also among these groups and national and international institutions, policies, and political actors.

Another dynamic of resource scarcity is the opposite of social scarcity. It might be characterized as social abundance. In such a situation, people deal with resource scarcity by increasing the number of social networks through which they acquire resource access. As modern institutions supplement or replace traditional institutions, a wider range of people control power resources and wealth-producing opportunities. In the case of forest resources, this may mean that more and more groups are making claims to all or part of the forest. Poorer people, with little direct access to such opportunities, are motivated to increase the number of social networks in which they are involved. By increasing their access to people with claims and access to forest resources, the poor increase the number of channels through which they can acquire forest resources. The more networks there are, and the more people are involved with multiple claimants to the forest, the greater the pressures on the resource itself (Berry 1989).

Resources and Property Rights

The multiplicity of resources with different natural characteristics in any agrarian system can also affect property relations, but ecological factors are frequently left out of any analysis of change in rights and relationships (compare Schroeder 1993). In tropical forests, the complications of studying property relations and farmer decision-making are compounded by the sheer number of species. Even if we limit our discussion to species that have economic, medicinal, or ritual value, the range and diversity of agroforestry products is immense, as are the tenure arrangements pertaining to them.

To take a single example, there are many ways to carve up the bundle of rights to a tree, as Fortmann (1985) has so aptly shown. The largest bundles include rights to own or inherit, to plant, to use, and to dispose of a tree. Each of these rights can be subdivided, and many of them interact with other factors. For example, tree tenure may determine or be determined by land tenure, or the two may be independent. Rights to use different parts of the tree may be allocated to different claimants, and these multiple uses may not always be compatible. Land rights and tree rights may be held by different claimants; change in one often, though not always, leads to change in the other. In addition, there are four broad classes of right holders: the state, nonstate social groups, households, and individuals; these too may be subdivided into different kinds of groups (by spatial units of residence, by kinship, by legal association, and so on). Whether a tree was planted or self-sown can make a difference in the claims of rights holders; whether it is used for subsistence or commercial purposes can also make a difference, according to Fortmann (figure 4.2).

One other influence on tree tenure bundles, to add to Fortmann's list, is a tree's natural characteristics: the length of time it grows and produces fruit, nuts, or other products and its reproductive strategies—how many seeds the mother tree produces, whether the fruit is harvested or left in the forest, seedling survival rate, and intensity of management (Peters 1994). All these factors influence people's ability to cultivate or manage a self-sown tree, as well as the tree's productive life span. Biological characteristics also play an important

Figure 4.2 Smallholder farmers weeding in a woodlot

role in inheritance patterns. Date palms in the Sudan, for example, can be inherited indefinitely by the original planter's descendants because new sprouts are continually emerging from mother trunks (Leach 1988).

State and International Influences on Tenure

State forest policy needs to be understood as a contemporary means of allocating access to resources—or property rights—which may change the distribution of benefits from forest use. Some common means of rearranging access include zoning land use areas, restricting access to particular forest species, and implementing controls on forest laborers. Although multiple-use management systems for sharing access to timber and nontimber products are theoretically possible, in practice they are often difficult to realize. When forest areas are set aside by governments for production or for conversion to large-scale agriculture (for example, plantations or agribusiness), local people's access to the forest products that sustain them is often seriously curtailed. It is impossible to understand why certain management techniques work or fail without understanding both the multiplicity of claims to the rights to manage and the variable foci of power to enforce the claims.

When a state takes over forest management and the administration of access rights, the nexus of rights enforcement is changed. Research has shown time and again that the movement of enforcement obligations and the distribution of benefits from local forest users to the state and to large corporations does not necessarily prevent the decline of forest quality and quantity (Fortmann 1990; Guha 1990; Peluso 1992b; Hafner 1990). This is true particularly where forest laws and policies are applied to vast areas and where the legitimacy of state claims is questioned by both local people and the local "keepers of the forest"—forest guards.

State systems of forest management arose largely in the eighteenth century in Europe (Meiggs 1982; Fernow 1911) and spread to other parts of the world during periods of European expansion, primarily in the eighteenth and nineteenth centuries. State forest management had varied degrees of effectiveness both within Europe and without, in areas under colonial rule, during the nineteenth century (Sahlins 1994; Hecht and Cockburn 1989; Grove and Anderson 1989). The most significant and lasting legacies of state regent systems initiated in this critical period were systems of law and policy that guided forest management practice. Even "noncolonized" countries such as Thailand were influenced by the tenets of European and colonial forest management because the king elicited advice from the British (Hafner 1990; Hirsch 1990).

Since the advent of independent administration of nation-states in the middle of the twentieth century, government claims to forest lands have increased. The culture of enforcement has become more technical, more bureaucratic, and more influenced by the military (Peluso 1993). Maps made increasingly accurate by modern geographic information systems technology, the increasing "scientization" of forest management methods, and the increasing power of forest bureaucracies in Southeast Asia have all become key components of forest management and of the justification of state claims (Peluso 1995).

International institutions are the newest actors to exert direct influence on local resource management. These include not only multinational firms, or "penetrating capital," but also international NGOs with resource protection or local participation mandates or goals and other international "movements" such as efforts to facilitate exchanging national debt for local nature preservation. Enforcement of international obligations regarding forests is a tenuous and complex matter. Because of sovereignty issues, international organizations generally cannot police claimed territories themselves; thus enforcement requires the cooperation of nation-state security forces. National and international claims on resources are increasingly constraining local people's resource management activities and rights (West and Brechin 1991; Peluso 1993) in favor of either national or "global" management strategies.

At the regional, national, and international levels of analysis, similar types of conflicts over claims of competing user groups need to be explored. Often groups with varying degrees of power or enforcement capability compete for the same resource. Their intended uses of the material have different potential effects on the resource and on the competing user groups. For example, national goals for generating revenues from forests may conflict with local people's in maintaining standing forests in order to collect nontimber products. Similarly, local people's interests in selective harvesting of their timber, hunting forest game, or using parts of the forest for swidden fields may conflict with some environmentalists' intentions to establish nature reserves. Other groups with claims to and interests in forests include local and international NGOs, bilateral and multilateral donors, multinational companies, politicians, and state agencies competing for control of the same or overlapping territories (Peluso, Turner, and Fortmann 1994).

In trying to understand the various types of social relationships within which forest users are embedded, policy makers and planners will have taken the first step in learning how to resolve conflict. On the other hand, by looking

only to human-environment interactions without understanding the larger contexts influencing an individual resource user's behavior, policy makers and analysts are bound
to fail in their attempts to stop degradation.

Simply understanding an ecosystem and its social relations is also not enough; events and processes taking place
in other areas can affect local dynamics. For example, a land-
slide or the establishment of a new plantation in a particular
watershed area may cause people to migrate to an adjacent
watershed, one that is already well populated. We cannot
understand the origins or causes of population pressure on
resources within the second watershed area if we do not understand the "push factors" sending people out of the first.
When more and more people crowd onto smaller and smaller
areas of land, it is frequently not only because they are re-
producing more rapidly but also because they have been denied access to larger areas of land. Conservation set-asides
such as national parks, nature reserves, protection forests,
and biosphere reserves have the same effect on local claims
to resource territories as do production set-asides such as
timber concessions, pulpwood plantations tracts, and plantation agribusinesses (tea, coffee, rubber, oil palm, cocoa, and
others). They also constrain local people's capacity to sustain themselves on the land or with products from the forest.

Unfortunately, with the recent excitement about global
resource management strategies we tend to ignore the practical difficulties of management at any scale. The wide variety of ecological, sociocultural, and political-economic circumstances, along with the globalization of resource access,
makes it imperative that local and layered constraints on
smallholders be better understood. The many layers of competing claims to forest resources everywhere strain local capacities for management. What is clear is a trend away from

grounding resource management decisions locally, even though local people's actions and activities in the forest remain the ultimate management focus. The professionalization of management, however, has not sufficiently ensured either the sustainability of forest resources or the well-being of forest-dependent people (Hirsch 1990).

Summary

- A forest is a social and political place.
- People's relations with other people are as important to understanding their use of the forest as are people's direct forest management activities.
- Poorer people with little direct access to wealth-producing opportunities can expand their social networks and thereby increase the number of channels through which they can acquire forest resources. The more networks and people involved, the greater the pressure on the forest.
- State forest policy and international influences need to be understood as contemporary means of allocating access to resources—or property rights—which may change the distribution of benefits from forest use.
- A political ecology approach to analyzing the ways forests sustain people's livelihoods and the ways these livelihoods are threatened can help identify critical processes, structures, and relationships affecting forest use.
- To do political ecology:
    1. Start by identifying resource users and particular resource management practices or strategies, institutions for resource access and control, and sources of conflict and cooperation.
    2. Examine relevant state and market systems or aspects of forest management superimposed on local or customary systems for the production of timber and other products.

3. Examine international influences and the reverberations of local activities outward to the national and international levels.
4. Assess unanticipated outcomes from the interactions between multiple resource strategies.
5. Think of local, state, and international systems of management as a set of cumulative effects on the ways individual people actually use forests.

CHAPTER FIVE

Lessons for the Future

CALESTOUS JUMA, TAPANI OKSANEN, AND
RALPH SCHMIDT

Whereas the total amount of forested area in the developed world appears to be rather stable as a whole or even to be increasing slightly, the situation is fundamentally different in most of the developing regions. These areas grew in population by about two billion, roughly doubling, between 1960 and 1990. They are expected to grow further by about three billion, from four to seven billion, between 1990 and 2020. As a result, the risk of continuing deforestation in the developing world is quite high (FAO 1995b).

A few countries contain most of the world's forests. The global area of closed forests (those with continuous tree canopies) is 2,400 million hectares. Eight countries contain two-thirds of these forests. Twenty-two countries account for 85 percent of the world's closed forests. Figure 5.1 shows countries containing one hundred thousand, five hundred thousand, and one million hectares of closed forest.

The great majority of countries contain forests that are important both for domestic sustainable development and for the global environment. But there is considerable variation among regions in the extent to which they are forested. Some

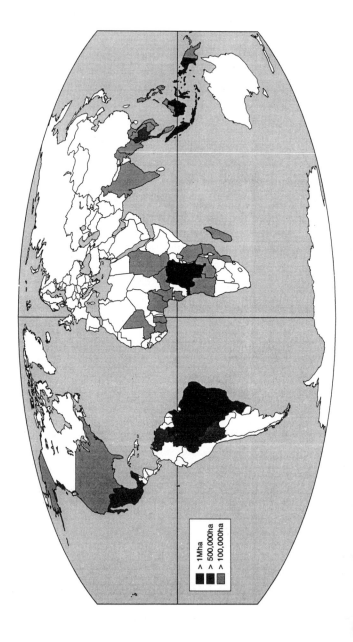

Figure 5.1 Map 1. Countries containing one hundred thousand, five hundred thousand, and one million hectares of closed forest, respectively

regions, such as the Sahel and southern Africa, had little aboriginal forest and have little today. Other regions were once
densely covered but have little remaining forest. West Africa,
for example, originally almost all forested, has lost more than
90 percent of its coverage and today is only 7 percent
forested. Central America (including Mexico) is only 10 percent forested, and further deforestation is proceeding rapidly
in both of these subregions. Only three regions are at least
one-third covered by closed forests: tropical South America,
central Africa, and insular southeast Asia.

Deforestation: A Permanent Loss

It is not surprising that the most forested regions are now
experiencing the greatest amounts of deforestation. Table 5.1
shows that South America (primarily the Amazon), southeast Asia, and southern and central Africa account for nearly
80 percent of world deforestation.

The FAO's forest resource assessments (FAO 1995b) established that eleven million hectares per year were defor-

Table 5.1   Annual Deforestation in the 1980s by Region

	Millions of Hectares	Percentage of Global Total
South America	6.2	40
Southeast Asia	3.3	22
Southern Africa	1.5	10
Central Africa	1.1	7
Central America	1.1	7
Sahelian Africa	0.9	6
West Africa	0.6	4
South Asia	0.5	4

Source: "Decline and dieback of trees and forests: A global overview."
Forestry Papers 120. Rome: FAO, 1994.

Figure 5.2 Map 2. Countries with the greatest amount of total deforestation

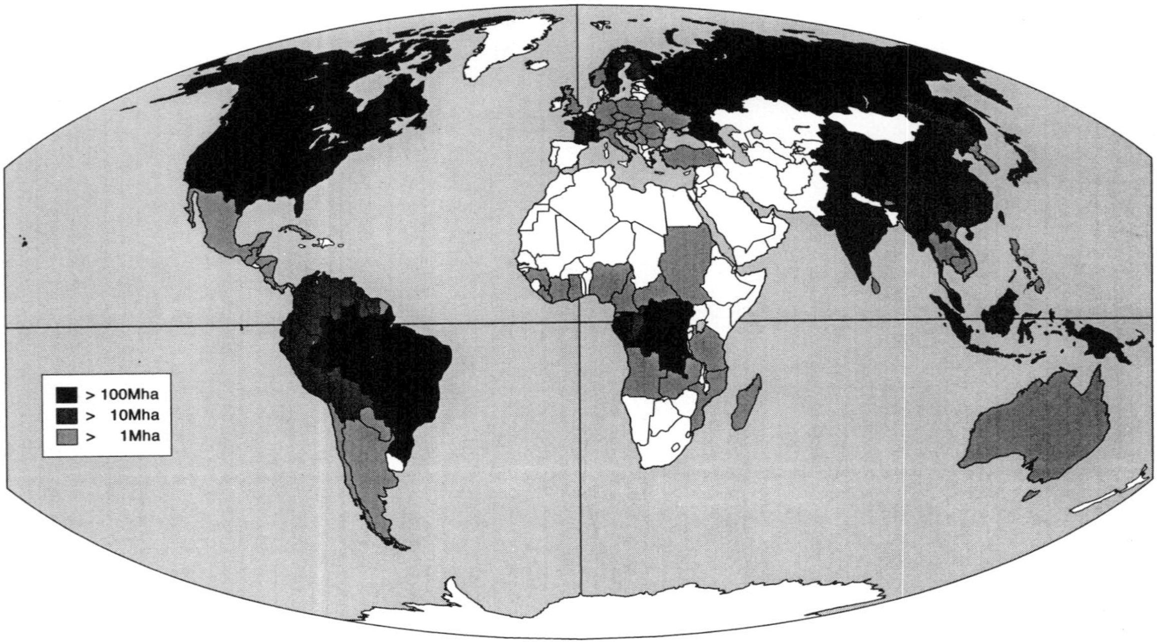

> 100Mha
> 10Mha
> 1Mha

ested in the 1970s, rising to sixteen million hectares per year in the 1980s. This amounts to a loss during these decades of a combined area the size of Peru and Ecuador. On the human time scale, these are permanent changes of global significance.

Deforestation occurs in many countries. But a relatively small number of countries account for most of the world's deforestation and degradation (just as a relatively small number account for most of the world's remaining forests). Figure 5.2 shows countries with the greatest total amount of current deforestation.

It is also important to assess the proportion of remaining forest being converted in each nation. Thus Brazil, according to FAO estimates, lost about one-half of one percent of its forests per year in the 1980s in a process that would take almost two hundred years to eliminate the forests. Costa Rica's annual deforestation is less than one-fiftieth the area of Brazil's. Yet it loses about 3 percent of its forests annually, and it will take only thirty years to eliminate them at the current rate.

A growing number of countries have lost essentially all of their closed forests. Eighteen countries have lost more than 95 percent, and eleven more have lost 90 percent. In ten of these countries, severe natural resource degradation has been accompanied by civil strife and the presence of United Nations peacekeeping forces.

An Overview of National Forest Planning

Over the past two decades, various planning frameworks have been used by a large number of developing countries to formulate and implement sectoral development programs for forestry. From 1985 to the present, the most widely used such framework has been the Tropical Forestry Action Programme (TFAP). Within this framework, national forest programs (nfps) have reached some stage of development in about ninety countries. Forestry master plans (FMPs)—promoted especially by the Asian Development Bank—and World Bank Forest Sector Reviews have also been widely used.

In addition to these programs, there are broader environmental planning frameworks and still more comprehensive frameworks dealing with national resource planning and management, such as national sustainable development strategies and the UNDP Capacity 21 program.

These programs were promoted by UNCED's global accord, Agenda 21, Chapter 11 of which stresses the need for intersectoral and holistic approaches to forest sector development. In order to combat deforestation, Chapter 11 specifically requests the signatories "to prepare and implement, as appropriate, national forestry action programs and/or plans for the management, conservation and sustainable development of forests" that "would be integrated with other land uses." Furthermore, the signatories agree to "reviewing and, if necessary, revising measures and programs relevant to all types of forests and vegetation, inclusive of other related lands and forest based resources, and relating them to other land uses and development policies and legislation" and "promoting adequate legislation and other measures as a basis against uncontrolled conversion (of forests) to other types of land uses." The challenge for forest planners and policy makers is to reconcile and make complementary sectoral and cross-sectoral programs while maintaining their compatibility with normal government planning structures and procedures and increasing their efficiency as frameworks for a coherent set of forest sector development programs.

Lessons from the Past

Despite the massive efforts of the past decades to arrest deforestation, forest degradation and the unsustainable conversion of forests to other forms of land use continue unabated, as shown above. The key constraints to implementation of forest sector programs at the national level are identified in several reviews and papers (see, for example, Guevara and Valle 1988; Hirsch 1990; IDRC 1991; Patel-Weynand 1996; Schmink and Wood 1987; Winterbottom 1990). These constraints seem to be present regardless of the sectoral planning framework used. Below, we summarize the most important constraints, discuss the principal underlying factors, and suggest how these problems can be overcome in the future.

Among the principal constraints on the implementation of sectoral plans and programs often have been weak intersectoral coordination and problems of policy and institutional reforms within the sector. Policies related to land use are still not coordinated toward the common goal of sustainable development. Effective high-level mechanisms for intersectoral coordination are rare. In many countries the lack of progress and transparency in the policy making and institutional reform processes has become one of the major obstacles to effective forest programs and international financial support.

These constraints, often labeled by the donor community as "lack of political commitment," reflect conflicts of interest between various stakeholder groups and indicate the complexity of the demands on natural resources in a developing country. The underlying factors are essentially political: land tenure and use are related to ethnic and social conflicts of interest. Also, economic realities in many developing countries—such as the need to balance the budget, meet debt payments, or satisfy key constituencies—impel governments to forgo long-term benefits for short-term gains.

In many developing countries forests absorb the demographic pressure on existing agricultural land. Deforestation is therefore closely related to the use of this primary productive asset. It is naive to believe that a program dealing with forestland as if it were a marginal sector representing only a small percentage of the GDP could have a major impact on these macrolevel issues. The most important policies related to deforestation usually fall outside of the forestry sector and relate to population, agriculture, energy, and trade (figure 5.3).

If an nfp is to have an impact on these constraints, it

Figure 5.3   Land deforested for cultivation

needs to be adopted at the highest political level and be linked to broader development planning. Those in an nfp process need to develop politically compelling arguments for sustaining forests and create a critical mass of well-informed and committed key decision makers. Experience indicates that politicians can be receptive to such programs in countries where the environment and the forests are becoming recognized concerns. Information-sharing also is essential in raising awareness. In order to facilitate policy and institutional reform, effective and transparent monitoring should be part of program implementation. Promising examples include national consultative forums on forests (for example, the Indonesian Consultative Group on Forests) and high-level interministerial coordinating bodies.

The prevalence of top-down planning and implementation and overreliance on the public sector in the implementation of productive activities are also common causes of failure in promoting forestry development. Despite a widespread understanding that successful forest interventions are demand-driven (based on local problems, priorities, resources, knowledge, and traditions) and rely on private initiative, too little has been achieved in making bottom-up planning and implementation truly operational. Project-level grassroots participation is increasing, but the experiences are rarely transferred to the national level. Also, the public sector often remains involved in profit-oriented production, which is better left to private enterprise.

The reliance on top-down approaches is principally caused by two deeply rooted traditions: (1) the centralized planning and autocratic governmental structures prevalent in many developing countries and (2) technocratic forest planning. Theoretical supply-demand gap models have been used (often with unreliable data), and, based on these results, programs have been designed to address estimated shortages of various products and services. These plans often have become extremely complex development blueprints in which the concept of local participation has been reduced to mobilizing people for some predetermined goals related to production and conservation. The models assume that the planners can design ideal solutions, but conflicts of interest between various stakeholders are not addressed. Generally, the local people have not been effectively involved, or even consulted, in defining program objectives, strategies, and approaches. As a result many forestry plans have remained relatively theoretical documents known in detail only to a limited number of experts, and national institutions and international and donor agencies pay little more than lip service to the plans.

The remedies for these problems include, first and foremost, focusing national-level planning on policy and institutional strategies that create an enabling environment for forest development at the local level. Decentralized forest development programs based on existing administrative units could be designed at the local level, using a variety of participatory and conflict resolution techniques. The public sector should perform a facilitating role, with implementation of specific activities left to the private sector, NGOs, and community-based organizations. Public-sector support of farmers, villages, and other private actors could be channeled through direct incentive systems such as subsidized grants and credits or through indirect systems such as extension and training or clarification and securing of land tenure.

An orientation toward attracting aid and a lack of clearly defined national priorities are other problems that often have diminished the effectiveness and sustainability of sectoral programs. Many of these have remained shopping lists of projects seeking external funding, without clearly defined priorities or linkages between programs. Too often national priorities have been set to attract external funding rather than focusing on constraints to local initiatives, incentives needed for local action, and cost-effective development assistance. In addition, many donor-supported programs have suffered from a lack of systematic and comprehensive capacity-building, which would enable governmental and nongovernmental organizations to assume proper responsibility for and leadership of sectoral development. As a consequence, dependence on external donors has remained high and the sustainability of the development activities low.

The lack of coordination of priorities between various national forest programs sometimes causes a perception within the donor community of inefficiency. When developing countries have set clear priorities, however, the donor community often has been reluctant to agree with them. Two sets of policies, priorities, and constituencies for the same program are thus created, the country's and the donor's, with each considering its interests perfectly legitimate.

The reasons for these problems are related to two features of official development cooperation that have proved difficult to change. First, most donor and financing agencies tend to apply their own priorities, procedures, and administrative practices without due regard for the negative impact of doing so on the growth of national capacity. This is exacerbated by the high reliance of financing agencies on external technical assistance personnel and by the agencies' short-term commitments and shifting priorities. Changes in development cooperation are often reactions to the demands of various interest groups in the donor countries. They are usually not due to changing priorities in the developing countries or improved understanding of the causes of the problems. Development cooperation organizations tend to apply their own perspective and follow their own priorities and principles in order to satisfy public opinion and meet legal requirements in industrialized countries.

This situation needs to be openly acknowledged and negotiations pursued in order to find mutually acceptable solutions. The matter is further complicated, however, by the fact that in many developing countries democratic institutions are still weak, and thus special arrangements may be required to ensure that priorities are defined based on the interests of the whole society and with due respect to the interests of minority groups.

Second, capacity-building often is ineffective owing to an outmoded definition of the role of government institutions in the development process—an increase in capacity would have greater impact in the private sector and with NGOs directly related to production. Examples of this problem include reluctance to transfer forest utilization, processing, and marketing to the private sector and inability to make full use of the capacity of local NGOs and CBOs. The donor and the international communities share the blame for the failures in capacity-building, which for more than three decades has been one of the principal objectives of development cooperation. A major factor undermining local capacity-building is many donors' and financing agencies' frequent bypassing of regular administrative structures. In doing so the donors and agencies do not enhance local capacity-building; they merely create parallel structures that attract the most qualified national professionals.

Program improvement is based on jointly agreed-on frameworks to enhance funding and cooperation. Concrete examples of this include bilateral and multilateral co-financing of national facilities and funds for sustaining forests. Other types of arrangements include national environmental funds (for conservation and reforestation, for example), which are being developed in several countries using a combination of debt-swaps and donations. International support for these arrangements depends on national institutions' ensuring transparency and accountability. Experience indicates that this is no simple task in many countries owing to the difficult conditions in which the local institutions and staff operate.

Aid funds are and always will be subject to changing trends and availability. In the final analysis, however, effective nfps will be based on national priorities and resources, complemented by strong international partnerships for cooperation.

Managing the World's Forests

A wide range of measures are being put in place to manage the world's forests. These include global and regional agreements, national policies and laws, and building of institutions. In some countries there is increasing funding for and emphasis on research and development in forest management. Despite the advent of various management techniques the world continues to lose a considerable percentage of its forests. Current efforts have potential for solving the problem, but they are disjointed and often unfocused.

Most of the work on forest management has approached depletion from a technical standpoint without examining institutional and policy facets of the problem on a global or an international scale. But effective forest management will only be attained if appropriate institutional structures and policies are put in place at all levels of governance.

An assessment of international, regional, national, and local institutional responses to the problem of increasing forest depletion can begin to remedy these deficiencies. Here, we examine institutional capacity-building in developing countries. We argue that in order to effectively manage forests for sustainable development, countries need to create new and focused institutional structures and accumulate capabilities at the national and local levels. Such measures should explicitly involve and support local community institutions. We suggest a number of policy and institutional measures designed to promote capacity-building for forest management in the developing countries.

Concerns over the increasing depletion of forests have their roots in the early 1940s. The work of the World Conservation Union/International Union for the Conservation of Nature (IUCN) gave institutional backing to these concerns. The United Nations Scientific Conference on the Conservation and Utilisation of Resources in 1949 gave the IUCN efforts international legitimacy. The conference was organized on the premise that the world's heritage of fauna and flora was increasingly being pushed toward extinction by man's socioeconomic activities. These concerns were later articulated at the United Nations Conference on the Human Environment held in Stockholm in 1972. Principle 2 of the Declaration of the Stockholm Conference called on nations to identify and institute international programs to manage forests and biodiversity. It stressed that "the natural resources of the earth, including the air, water, land, flora and fauna, and especially representative samples of national ecosystems, must be safeguarded for the benefit of present and future generations, through careful planning or management, as appropriate."

The Stockholm Conference provided an impetus for growing concerns about forest management. In 1982 FAO, in collaboration with UNEP, published the most comprehensive assessment to that date of tropical forest resources. It was alarming to realize that at least eleven million hectares of tropical forest were disappearing each year. The World Commission on Environment and Development (WCED), established in 1983 by the United Nations General Assembly and headed by the prime minister of Norway, Gro Harlem Brundtland, made major efforts to bring the urgency of identifying and instituting long-term measures to conserve the environment to the center of the international stage. The WCED was charged with identifying long-term strategies for achieving sustainable development through environmentally sound means and with suggesting legal, policy, and institutional mechanisms that would ensure that nations act together in dealing with global environmental and development problems. Although it addressed a wide range of issues, from climate change to soil erosion, the WCED gave significant attention to forest management. The WCED brought to the mainstream of international policy issues of forest management that had largely been confined to technical and scientific discussions.

The commission recommended to the UN General Assembly that an international conference on environment and development be held in 1992 to evaluate progress in dealing with environmental problems since the Stockholm Conference. Since the report of the commission was published in 1987, forest management has become one of the top issues on the international environmental agenda. A number of international, regional, and national initiatives have evolved to address the issue. On the international scene, forest management has taken a significant place in the United Nations' and other international organizations' activities on environmental management. International institutions such as the FAO, the IUCN, the World Resources Institute (WRI), the UNEP, the UNDP, the World Bank, and others have established programs on forest management. International leadership and institutional collaboration have been weak, however (Schmidt 1993).

Forest management received intense attention during the preparatory process for the United Nations Conference on Environment and Development (UNCED). Some of the major products of UNCED are the Convention on Biological Diversity, Agenda 21, and the Forest Principles. The Convention on Biological Diversity is an important instrument for promoting the management of forests. The objectives of the convention are "the conservation of biological diversity, the sustainable utilization of its components and the fair and equitable distribution of the benefits arising out of the utilization of genetic resources, including by appropriate access to genetic resources and by appropriate transfer of relevant technologies . . . and by appropriate funding." The convention was signed by 157 governments at UNCED.

The convention has provisions that require signatory countries to formulate measures aimed at enhancing national capacities for biodiversity management. Article 12 provides that the contracting parties shall

   (a) Establish and maintain programs for scientific and technical education and training in measures for the identification, conservation and sustainable use of biological diversity and its components and provide support for such education and training for specific needs of developing countries;
   (b) Promote and encourage research which contributes

to the conservation and sustainable use of biological diversity, particularly in developing countries.

Article 18 calls for establishing and strengthening technological capabilities for research and conservation in developing nations through international cooperation, stating, "In promoting such cooperation special attention should be given to the development and strengthening of national capabilities, by means of human resource development and institution building." The convention provides a framework that should relate coherently to programs for forest management. There is more to biodiversity than forests, however, and there is more to forest management than biodiversity.

The Forest Principles represent current international consensus on the management of forest resources. The principles agreed on by governments during UNCED include the following:

a. Forests are essential to economic development and the maintenance of all life forms;
b. National forest policies should recognize and duly support the identity, culture, and rights of indigenous people, their communities, and all forest dwellers; and
c. All aspects of environmental management, including those of forest protection, should be integrated with social and economic development.

These are essentially guiding principles. They are not legally binding. Governments can, however, apply them to develop specific policies and programs aimed at enhancing the management of forests.

Another important international instrument that can be applied to promote the management of forests is Agenda 21, which was adopted by governments at UNCED. Chapter 11 of Agenda 21 deals with combating deforestation. It recommends a number of measures for building and utilizing national capabilities for sustainable management of forests, and it calls on governments to undertake policy and institutional reforms to these ends. The chapter also provides a model program on capacity-building in the forest sector.

On the whole, the Biodiversity Convention, the Forest Principles, and Agenda 21 have set basic guidelines and measures that countries can use to establish programs for applying technology in the conservation of biodiversity. They have also stressed the importance of building national technological capabilities for undertaking research on forests.

International programs aimed at managing forests have also been established. One of the earliest was UNESCO's Man and Biosphere Programme (MABP), launched in 1971. The program focuses on the conservation of natural areas and forest ecosystems. It specifically deals with identifying habitats where species are threatened through research and field assessments, as well as identifying needs in training and building capacity for scientific research. By 1988, 267 reserves had been set up under the program.

Other international initiatives include the International Union of Forestry Research Organizations (IUFRO), which aims at enhancing collaboration among forestry research institutions. It facilitates the creation of partnerships among institutions and organizes international conferences to promote the sharing of information on forestry research. The organization is also actively involved in promoting training of forestry researchers. The Commission on Sustainable Development (CSD) is engaged in international reviews of forest status, policies, and practices through the Intergovernmental Panel on Forests.

Another international initiative that has bearing on

forests is the Global Environment Facility (GEF). The GEF provides financial support to projects that benefit the global environment. It is implemented through three agencies: the UNDP, the UNEP, and the World Bank. It focuses on four major global environmental issues: climate change, biological diversity, ozone depletion, and international waters. The GEF has provided support to a number of projects on forest management. In the three-year pilot phase of the GEF, $275 million was programmed for forestry-related activities. Projects supported by the GEF must feature use of appropriate technology, cost effectiveness, benefits for the global environment, and consistency with existing environmental conventions and national environmental strategies. On the whole, the GEF aims at providing financial and technical support to developing countries to build up technological and institutional capacities.

Examples of National Institutional and Policy Regimes

Different countries have established national programs and institutional regimes for forestry research and management with varying degrees of congruence with international initiatives. Countries obviously differ in their technological capabilities to deploy management techniques and conduct research aimed at conserving and sustainably using forests. They also have different institutional regimes with varied technological abilities. In addition, various nations have established a wide range of legal and policy measures for promoting forest management. Furthermore, the philosophical underpinnings of forest management programs have tended to vary, though there are considerable similarities in the approaches used.

In Kenya, for example, the management of forests is largely influenced by colonial policies and by institutional orientation toward preserving forests in reserves. This approach—which also prevails in many other countries—has its origins in the colonial systems of forest management. The emphasis under the colonial approach was on preserving forests in areas not interfered with by humans and their socioeconomic activities. In Africa particularly, forming reserves is the most common approach to protecting and managing forests. As described in Chapter 3, however, most forests in Africa and elsewhere have long been under some form of human management.

In Cameroon, emphasis has been placed on the conservation of fauna and flora through the protection of vast areas of forest. Several national forest areas have been established in the northern, southern, and central parts of the country during the past twenty-five years. The areas are protected from extractive activities—essentially deforestation—by fencing and policing.

Kenya has one of the most elaborate institutional forestry research and management systems. The national and international institutions engaged in forestry research and management include the Kenya Forestry Research Institute (KEFRI), the Forest Department, and the International Centre for Research in Agroforestry (ICRAF). Other institutions, such as the Department of Resources, Surveys, and Remote Sensing (KREMU), have undertaken activities aimed at assessing the status of forests. The country has also a number of NGOs that deal with various aspects of forestry research and management. Notable examples include the Kenya Energy and Environment Nongovernmental Organizations (KENGO), the Forestry Network, and the African Centre for Technology Studies (ACTS).There are a number of community-based projects supported by KENGO that focus on various aspects of forestry management, whereas ACTS is engaged in policy research projects that focus on land tenure and forest policy.

On the whole, Kenya has the basic institutional setup required to ensure effective forest management. But it lacks the institutional synergy and focus needed to manage its forests properly. It needs to identify ways of establishing institutional linkages and strengthening the capacity of the existing institutions to work together. In addition, the Forest Department, which has legal oversight of the management of forests, has very limited capacities, particularly in research and management. More than 60 percent of the department's technical staff are forest wardens with limited training in research and little knowledge of new technologies for forest multiplication and management. Furthermore, the annual budget of this department has declined by more than 25 percent during the past decade or so, when the demand for forest protection and management has in fact grown.

In Ghana, the Forest Department, created in the early 1900s, is charged with overseeing the management of forests. It is mandated to ensure effective forest management as well as the implementation of a national forest policy that was first put in place in 1940 and reformed in 1989. The policy stresses (1) the need to establish reserves as well as promote a forest station; (2) increasing the application of science and technology in the management of forests; and (3) promotion of sustainable utilization of forest resources. The Forest Department is responsible for overseeing the implementation of these elements of the policy. The capacity of the department to implement the policy as well as enforce various regulations controlling the overexploitation of forests is limited, however. A study conducted in 1989 with the support of the British ODA demonstrated that the department lacks adequate trained personnel in strategic areas. It noted that more than 75 percent of the department's staff members are forestry officers trained in traditional forest protection within reserves. Most of them are holders of undergraduate degrees and certificates. The capacity of the department to engage in scientific research involving the application of new technologies is quite low.

On the whole, forest managers in developing countries face problems that limit their ability to carry out their mandates. These difficulties include low levels of funding for research and management, poor infrastructure, lack of skilled personnel, conflicts in land-use systems and policies, lack of scientific information on the status of forests, and lack of strong science- and technology-based institutional setups for managing the forests. In many developing countries the institutional and legal regimes for forest management are characterized by jurisdictional and technical inadequacies. In Zimbabwe, for example, a number of institutions have been created to work on various aspects of forest management. They are often locked in conflicts regarding jurisdictional mandates and sectoral administrative authority, however. Some of the institutions are placed in bureaucratic systems that deny them the flexibility and autonomy required for change and growth. They often have to compete with conflicting sectoral administrations, such as that for agriculture, that are expanding at a rapid rate. These institutional conflicts and inadequacies have often resulted in problems of coordination and management and have limited the technological capabilities of countries to manage forests.

As discussed in the section on international cooperation, decades' worth of donor involvement (and often control), as well as substantial funding to strengthen national forestry institutions, have not produced sustained improvement and the required capacities for forest management. Donors have tended to support isolated projects rather than considering the economic and institutional contexts in which the projects are implemented. In many cases they have promoted institutional conflicts and nurtured sectoral approaches.

The management of forests is usually in the hands of the state and a relatively small number of individuals. The public in general has rarely been involved or consulted on how they wish forest resources to be used. The management and control of forests have been seen as the prerogatives of public institutions and trained foresters within the institutions and have until recently seldom been discussed as public policy issues. Forest policies have been formulated without adequate consultation with the stakeholders. In some countries, specific laws and policies dealing with forestry management not only are fragmented but date back one hundred years or more.

Some developing countries have made efforts to revise their forest policies. A notable example is India, which has invested considerable resources in the revision of its forest policies and laws to reflect the role of local communities in conservation. Other countries, particularly those of Africa, have been unable to engage in forest policy reform. There are a number of reasons for this. First, the nature of forestry institutions in these countries has hindered change. The institutions that were created decades ago are rigid and resistant to change. Second, most of these countries lack adequate policymaking capacities. Third, there is a lack of genuine governmental support for policy reform. Some senior governmental officials and institutions have major interests in continuing to control and exploit forests without regulation.

In most African countries where forest policies exist, the

policies are often characterized by vagueness and confusion. They have not evolved in response to national needs and socioeconomic changes. Instead they were devised by governments, sometimes in response to external demands and often to deal with short-term crises. These policies have short-sighted, simple, and single objectives that reflect the interests of government rather than those of all the communities and institutions that have a stake in the use and management of forests. Furthermore, in some countries there is a disjunction between stated policy and law and practice. The existing forestry institutions are unable to implement the policy and enforce the law, which are really window dressing. Forest destruction continues despite their existence.

It is vital to note that although policies, laws, and plans are necessary for effective management of forest resources, their mere existence is not a sufficient condition for that management. The increasing internationalization of forest management and growth in the global forestry industry mean that nationally focused policies and plans become inadequate in scope. In addition, major economic and trade policy changes that affect the management of forests are taking place. But in many cases there is a lack of integration of forest policy with other policies dealing with industry, agriculture, tourism, and environment. Given the importance of these sectors in global and national economies, their policies have greater effect on the use and management of forest than have forest policies. The point is, forest policies should not be formulated in isolation from the broader concerns of economic change and sustainable development. The policies need to take into account not only national socioeconomic needs but also global changes.

The challenge for most developing countries is, therefore, to formulate integrated forest policies and plans. As discussed in Chapter 7, these must reflect the increasing role of local communities, the private sector, and international agreements and institutions in the management of sustainable development processes. In some countries of Africa the formulation and implementation of forest policies and plans demand the existence of supportive land-use policy and management systems; forest management is just one of an array of land-use activities. In the absence of a coherent land-use policy—one that aims at promoting the management of natural resources on the land as well as using the land as an important economic asset—it is difficult to envisage effective forest management. Furthermore, policies for other sectors are vital in dealing with forest issues. It is critical that institutions and policy makers dealing with other sectors such as agriculture, tourism, and industry take into account both the direct and indirect effects of these sectors' policies on forests and carry out research in more narrowly defined fields. More of this expertise is within local community institutions. The outcome has often been the underutilization of expertise and equipment, lack of knowledge of the national requirements for effective forest management programs, and failure to attract support for these activities.

Community-Based Forest Management

The role of community-based institutions in forest management has recently been recognized. Traditional communities have cultural ties with their land and the natural resources on it. Through experimentation and learning, traditional communities have developed holistic approaches to addressing ecological problems. These approaches are largely a reflection of the various forms of social organizations that the traditional communities formed (figure 5.4).

It has become increasingly apparent that traditional institutions can be highly effective in sustainable forest management because they hold knowledge and skills appropriate for dealing with most of the ecological problems that confront mankind. This fact was recognized by UNCED, which incorporated traditional institutions and experiences into Agenda 21. Indeed, the failure of tenurial reform to promote sustainable forest management may be attributed directly to the inability of those reforms to incorporate traditional institutions into the framework of strategies. Such incorporation can, however, only be sustainable in the long run if founded on a clear understanding of the nature of the institutions and their place in the overall social system. In-

Figure 5.4   Government forestry officer explaining a community forestry development project to villagers

corporating them into tenurial reform programs will ensure that traditional tenure systems and institutions are responsive to changing circumstances and influences. It will also provide space and incentives for local people to participate in and invest in forest management.

Forests and forest resources are tied to the socioeconomic structures of most of the local communities in different countries of the world. The intimate link between traditional socioeconomic systems and forests is manifested in the various economic activities, rites, and rituals undertaken by the local communities, whose socioeconomic activities depend in large measure on the availability of a wide range of plant and animal resources, most of which (including medicines) are obtained from forests. Furthermore, certain species of trees hold particular spiritual value for forest dwellers. The loss of woodlands, therefore, means the erosion of the cultural and economic bases on which the local community depends.

Local communities in different parts of the developing countries have developed various measures for managing forests. Most of these measures are to a large extent related to the nature of traditional tenure systems. Some of the communities' tenure regimes provide for ownership of specific types of trees. For example, recent ethnobotanical studies demonstrate that the Turkana living in the arid areas of northern Kenya have tree rights. This is contrary to the established view that the Turkana are essentially pastoral people whose socioeconomic systems are entirely dependent on the genetic resources held in their livestock. The ownership of trees among the Turkana brings with it specific responsibilities of managing the trees and their ecosystem.

In general, local communities have also consolidated a broad base of knowledge for managing forests and forest resources, and this knowledge is, to a large extent, reflected in the nature of community institutions.

Most national policies still embody colonial systems that perceived traditional communities as primitive, with no knowledge of ecological management. These policies and laws have been enforced by rigid public forestry institutions, most created by the colonialists, that ignore local knowledge and management systems. In many cases the intimate links between the communities and their socioecological fabric have been destroyed. This has had dire consequences for the management of forests.

A number of developing countries have initiated programs on community forest management. Most of these programs have been founded on the increasing recognition that while governments have over the years tried to expand their control of forests and have to a large extent created systems that destroyed the socioecological fabric of local communities, governments have themselves failed to ensure effective forest management. More and more local people have taken charge of the democratization process to demand the restoration of their tenure over forests and forested lands. Restrictive laws and regulations governing land ownership and use have been found to be ineffective. In many cases these have created disincentives for community investment in forest management. Some governments, such as India's, have started reforming their resource tenure regimes to respond to the need to bring local communities to the center of resource management.

In India, the government has undertaken major policy, legal, and institutional reforms aimed at strengthening the participation of local communities in national forestry activities. In June 1990 the government of India issued a set of guidelines aimed at enabling various state governments to encourage local communities and nongovernmental agencies to engage in forest management. A number of states (for example, Jammu, Haryana, Bihar, and Gujarat), have adopted the guidelines and initiated what are called joint forest management programs. The guidelines enable the states to reform their policies and laws on forest protection to allow community tenure of resources. In some cases, as in West Bengal, the communities are allowed to participate in the protection of local forests as well as the use of forest resources. The people are allowed to use all nontimber forest products and receive 25 percent of revenue generated from timber sales. Local people are beginning to form social organizations for the purpose of managing the forests. The Forest Department is also working with the local groups in various activities of forest management including reforestation. Partnerships are emerging between local groups, state government authorities, and the Forest Department.

Although the case of India demonstrates an emerging paradigm shift in the institutional approaches to the management of forests, it should be noted that most developing countries still maintain excessive restrictions on the participation of local people in forest management. Most of these restrictions are in the form of prevailing resource tenure regimes, as well as rigid forestry institutions. There are a number of institutional issues that limit the participation of local communities in forest management. First, national forestry institutions are larger and stronger than local ones. Most still apply traditional management approaches that in fact alienate local people from forests and deny them access to forest resources. More generally, local people are not likely to benefit fully from the forests until new legal and institutional regimes that provide space for their participation are established. Accordingly, most developing countries should undertake deliberate but careful policy and institutional reforms in order to address factors that are likely to hinder the participation of local people in forest management. Most of these reforms should be undertaken within the framework of national environmental policies and legislation. In other cases the reforms should be reflected in national constitutions so that issues of resource tenure and local people's rights are firmly grounded within overall governance.

Second, some countries have limited national capacity for undertaking policy and institutional reforms. In such cases, it is important that support be first directed to the creation and enhancement of national policy formulation and management capacities. This is an area where multilateral bodies may play a role. Currently, some countries are simply establishing isolated programs on community forestry management. Most of these efforts are not focused within the context of specific needs for national institutional reforms. What is likely to occur after a time, if such reforms are not undertaken, is that tension between the community programs and the traditional institutional approaches will emerge.

Public Policy Issues

The need for countries to undertake policy reforms has been articulated in Chapter 11 of Agenda 21, which notes that "there are major weaknesses in the policies, methods and mechanisms adopted to support and develop the multiple ecological, economic, social and cultural roles of trees, forests and forest lands. . . . More effective measures and approaches are often required at the national level to improve and harmonize policy formulation, planning and programming, legislative measures and instruments, development patterns, involvement of youth, roles of the private sector, local organizations, nongovernmental organizations and co-operatives. . . . This is especially important to ensure a rational and holistic approach to the sustainable and environmentally sound development of forests." The chapter goes on to offer a wide range of areas of policy reform and stresses the need to build capacities for policy reform and management.

On the whole, the strengthening of institutions in developing countries should be accompanied by specific policy reforms, which need to cover land-use planning, land tenure, and economic development in general. It is on the basis of such reforms that suitable policy environments will be created to ensure effective forest management under conditions of well-developed land-use strategies, unambiguous resource rights, and economic planning sensitive to environmental concerns.

The importance of international cooperation in promoting forest management has also been recognized in Chapter 11, which stresses the need for such cooperation to facilitate the sharing of expertise, resources, and information. There are numerous opportunities for partnership in forest management. These include joint research programs, exchange of scientists, and long-term institutional collaboration.

International collaboration is critical particularly when considering transborder forests. Given the increasing global concern for the fate of forests, the Commission on Sustainable Development in April 1995 established the Intergovernmental Panel on Forests. If it and similar efforts are to succeed, governments must learn to work together. The recently completed work of the IPF is evidence of movement in this direction.

Ultimate success in forest management will largely depend, however, on how the developing countries organize their institutions and build the requisite capacities at the national and local levels. This is mainly because forest managers increasingly require a wide range of scientific, managerial, and technical expertise.

Summary

Rapid conversion of forestland is occurring in almost all developing countries. A growing number of developing countries, especially in Central America, the Caribbean, western and southern Africa, and mainland Asia, have lost virtually all of their old-growth forests. Relatively few countries (ten to twenty) account for the great majority of forest area and of current deforestation.

A considerable amount of national forest planning, programming, and project implementation has occurred, although the scale is still minor compared to the economic and human population forces involved. Lack of acceptable results may be attributed to this scale, but so may the real conflicts of interest within affected countries and among forested countries and rich nations that contribute to international cooperation. New forest programs will face the daunting challenge of requiring strong high-level political support and broad participation on the part of the poor. In many countries, international cooperation is essential to confront the problems, but it is managed in ways that are totally inadequate for successful outcomes.

International action on forests takes place within a thirty-year-old system of international agreements on global environment and development issues and institutions to support agreed-on actions. The major agreements are reviewed, and national examples of implementation are considered. Extremely antiquated forms of administering forest resources, with the state's taking on roles it cannot fulfill, have persisted in many developing countries. Forest policy can be no more coherent or effective than land-use policy overall. Many of the most promising developments occur when forest management is placed in the hands of the local community. This points the way but still requires that developing countries organize their institutions and build capacities for the wide-ranging demands of sustainable forest management.

CHAPTER SIX

Human Resources Development

RUBÉN GUEVARA

Knowledge itself is Power.
—FRANCIS BACON

Most conventional assumptions about power in Western culture at least imply that power is a matter of quantity. Although some of us clearly have less power than others, however, this approach ignores what may now be the most
important factor of all: the quality of power. The highest-
quality power comes from the application of knowledge.
Therefore, it is not simply clout. High quality implies much
more: efficiency in the use of knowledge (Toffler 1990).

Knowledge itself turns out to be not only the source of the
highest-quality power but also the most important ingredient of force and wealth. Put differently, knowledge has gone
from being an adjunct of money power, and muscle power,
to being their very essence. Knowing this precept, the director general of UNESCO, Federico Mayor Saragoza, recently
said: "Knowledge is power, and people can use it to the extent they know," adding, "the destiny of countries is forged
by the talent of their citizens, without regard to the size of
their population or territory" (Saragoza 1995). Not coinci-
dentally, the decisive factors in sustainable development are
improvement in population quality and advances in knowl-

edge. Education accounts for much of the improvement in population quality. Perhaps the most vital difference between developed and developing, rich and poor, is the knowledge gap—disparity in the capacity to generate, acquire, disseminate, and use scientific and technological knowledge. The extent of this capacity will make the difference between the parts of the world where people are able to decide and act independently and those where they cannot (IDRC 1991).

The empowerment of people, then, comes through the various forms of education, including the legitimization of local knowledge, and through participation. *Local knowledge* conveyed a pejorative or derogatory impression in the past. A growing number of development experts have changed that view, however, and today the term is positive and laudatory in most respects. Most of this knowledge is the by-product of experience and empirical experimentation through several generations, which may be strengthened by cultural traditions that have co-evolved with the local environment. In some cases, the knowledge is based on unique epistemologies, philosophies, intuition, and principles, and it may differ from modern scientific tenets. But it is proved that its importance is beyond any doubt (Thrupp 1989).

Empowerment though participation involves the recognition that local people, through their intelligence and personal experiences, are suited to decide what is best for their development. This process recognizes that true participation occurs when decisions by government, services provided by the state, control of external productive resources, and priority setting are carried out in conjunction with the beneficiaries of these actions. Full participation is achieved when there is the political will to enforce such policies.

The UNDP in 1994 stated that "development will not

merely come from introducing better crops, new cattle breeds, more credit or rural cooperatives, as important as these may be. Rather, it is achieved by farmers working in very specific farm-household systems. It must be based on the tasks, needs, and aspirations of the farmers themselves and on the dynamics and constraints they face, not only in their farming but also in the domestic and non-farm activities. It must take account all of their rural life situation, including real-wealth factors beyond the control of the household: the ecology and natural resources of the zone, the social-cultural environment in the community, and the policies, prices, services, and infrastructure that affect rural prospects" (UNDP 1994b). Therefore, in order to foster participation, programs should encourage people to reflect their own conditions, speak their language, consider their locality's needs, and take their interests and values into account. Furthermore, the process has to respect them as individuals, find ways to get them to have a stake in what they do and receive direct benefits.

In many cases participation is seen as difficult, time-consuming, and tricky because of the probability that small elites of free riders can get more than their share and because this may stir up conflicts that traditional societies and cultures have been able to keep under wraps. James Gustave Speth explained the importance of participation in sustainable development, however, in his address to the UNDP staff in July 1993: "Sustainable Human Development is participatory. It can only be achieved when people have an opportunity to participate in the events and processes that shape their lives; where entrepreneurs, women, non-governmental organizations, and others in civil society are empowered to take initiative and participate in both open markets and effective governments, and where pluralism prevails and human rights and access of information to all parties are guaranteed." He also emphasized the need to pursue this new understanding of development in today's world, which will need a new approach to international cooperation, with key elements including the building of initiatives from the bottom up, ensuring ownership, and working with new groups that have been neglected in the development process: NGOs, women, entrepreneurs, indigenous people, participants in the informal sectors, and local communities (figure 6.1).

This participation, with the appropriate knowledge, should be the basis for development. It is a process by which

Figure 6.1 Multidisciplinary approach to education, research, and outreach

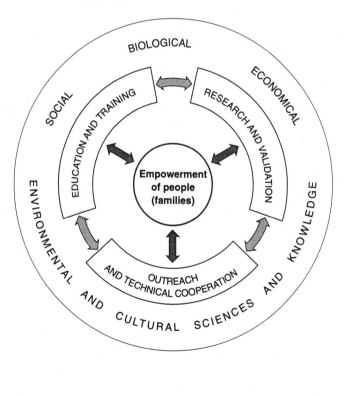

people influence the things that affect them. If they participate they therefore have a chance to become the masters of
their own destiny.

The new paradigm of sustainable development, in which
forestry is included, implies environmental protection, participation, natural resource amelioration, and equity. This
entails the development of a sustainable society, with a new
ethical frame of reference for those involved so that they
adopt a positive attitude and patterns of behavior in keeping
with the adequate use of natural resources. This can only be
achieved through education, research, and outreach.

New knowledge addressing sustainability has to be developed and disseminated quickly, so as to produce a multi-
plier effect leading to the achievement of the new paradigm.
This new way of development is people-centered, environmentally sound, and highly participatory if it is built on
local and national capacity for self-reliance. It is based on
sustainable human development with creativity, entrepreneurial savvy, and equity. It always addresses socioeconom-
ical, biological, environmental, and cultural concerns in a
multidisciplinary, holistic manner. This is the challenge for
impact assessment of international forestry research: to ask
how the output of such research contributes to meeting today's needs (for example, its contribution to income) and
how it may affect the capacity of future generations to meet
their needs (for example, its contribution to sustainability).

Sustainable Forest Development and Conservation

Various fora and the research community have concluded that in the tropics, the present model of forestry development is unsustainable because it leads to the inappro-

priate use of forest resources. Inappropriate use is caused mostly by the extreme poverty, which is originated by inadequate policies and legislation on macroeconomic and sectoral matters pertaining to land use; structural and institutional deficiencies or weaknesses; an imbalance in the different kinds of training and education; the lack of education generally; the lack of research and validation in sustainable forestry development and conservation; the inefficient outreach and technical cooperation provided to people in communities inside or around forests; and the inappropriate empowerment and participatory approaches of the local communities in the development process.

The root problems in forestry development and conservation can be addressed if (1) education and training are improved, (2) research and validation are augmented and refocused toward addressing sustainability, (3) validated research results are disseminated through an articulated outreach and technical cooperation program to the people with the most need, and (4) these activities are supplemented by appropriate participation and empowerment methodologies, proper policies, and legislation for land use. With these mechanisms in place the alleviation of poverty and therefore the appropriate use of forest resources would have higher chances of success.

In its expansion toward a better quality of life, sustainable development needs sufficient skilled and semiskilled labor. Many developing countries bet their entire economic future on the theory that selling labor cheap will lead to modernization. Cheap labor can be expensive, however, if it is not well trained and knowledgeable. The reason for this is simple: under the newly emerging system of wealth creation, unskilled labor is no longer enough to ensure market advantage in developing countries.

Education and Sustainable Forest Development

Figure 6.2 shows an ideal human resource development pyramid in which the base, representing the largest number of people, would be made up of skilled and semiskilled labor or the trained workforce. In the middle layers there would be a progressively smaller number of technicians and professionals coming from secondary education institutions, vocational or technical schools, and colleges and universities. At the apex there would be a relatively smaller number of people with postgraduate education at the master's and Ph.D. levels. The advancement in the degree of training would be higher at the apex of the pyramid, which would tend to make knowledge more sophisticated.

In many developing countries this ideal pyramid usually has a narrower base and a larger middle, the product of policies oriented toward supporting higher levels of education at the expense of training the workforce or the local forest dwellers. When the base of the pyramid is too narrow and the

Figure 6.2   Ideal human resource development pyramid

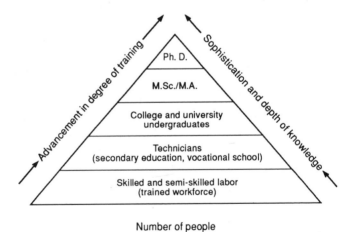

Number of people

middle too wide, the foundation on which it rests is unstable, leading to limitations in sustainable development.

In forestry development the pyramid must include extension workers, technicians, foresters, biologists, economists, sociologists, anthropologists, political scientists, and experts in community participation. At the bottom of the pyramid there must be a sufficient number of skilled and semiskilled workers incorporating external and internal knowledge from their places of origin.

The highest levels of education should be restructured to place more emphasis on fundamental research tools to enable future scientists to meet greater challenges. A larger number of women and members of minority groups must be recruited into forestry research, education, and outreach. In order to enhance the quality of forestry research, it should be opened up to a broader scientific community and increased participation by scientists from other aspects of life.

When education and training have been implemented properly, people will have real opportunities to develop their potential and to increase their self-confidence and desire for advancement. They will be aware that they themselves can solve most of their problems with the tools and the power they now possess to initiate their self-development. By then, they must be trained to identify the causes of their problems and to estimate their importance, and they should have the tools to eliminate or attenuate them. They should also be capable of correcting the external factors causing their problems; in other words, they should be able to discriminate between the important and the relevant and be capable doing what they can do, leaving aside all the tasks that they cannot do. This will lead to increasing their efficiency and utilizing their time wisely.

Education and training will also lead people to participate

more effectively in defining the needs for research and validation to solve their most common problems, which will bring the development and validation of science and technology with less risk, less energy and capital, positive ergonomic effects, and minimum dependence on outside sources.

Training is the external factor that has the merit of freeing the trainees from dependency on other external factors. There resides its extraordinary strategic importance, especially when those external factors are scarce, insufficient, or inaccessible. Training would also

1. Increase productivity in the members of the whole family: women, men and youngsters,
2. Stimulate a change in attitudes and values including self-confidence,
3. Widen the horizons of aspirations, increasing the desire for self-development, and
4. Consolidate the assumption of personal responsibilities both at the family and the communal levels.

Training can make local people think and behave entrepreneurially and therefore help them to emancipate themselves, making empowerment through knowledge a reality. Support for national research training should be encouraged since it has a number of advantages over training received abroad. To begin with, degree candidates tend to become engaged in research areas of direct importance to the country's development. While studying they can also assist university lecturers, who are usually in short supply, with the teaching of undergraduate students. Their research can serve as useful study material, which is an otherwise scarce resource in most developing countries. Furthermore, after obtaining their degrees, the vast majority of students remain in the country, contrary to the usual practice of researchers wholly trained abroad.

In true participation, the basic premise is that only rural families can promote their own development. Other agents or factors can only contribute to the family's own efforts. Training will lead rural families to have the desire to know and to be able to solve their own problems. Training should then focus on building endogenous capacity in the individual, in the community, in the region, and in the country.

According to T. Schultz (1979), education should not be seen as solely a consumer good but rather as an expenditure that will produce returns, and an integral part of the modernization of the economy of any country—and education's relation to the rise in human capital—has to do with adequate skills and knowledge. He adds that investments in improving a population's quality through education can significantly enhance the economic prospects and welfare of the poor.

All countries having forest resources of any significance should, as a policy matter, develop a cadre of forest scientists as well as scientists and educators in related fields that would constitute a support base for national research and education and on whose knowledge and experiences would be based all formulation of policy and legislation dealing with forestry in that country. Recruitment efforts for women and members of minority groups should be directed toward high schools, undergraduate technical schools, and postgraduate programs.

Agenda 21 addresses the issue of education and training, emphasizing the link between human activities and environmental protection. It states that formal and informal education can foster ecological and ethical behavior and that strengthened attitudes and values can lead to responsible behavior aimed at sustainable development.

Training and education should also include socioeconomic and cultural aspects of development. Agenda 21 stresses that all countries should implement policies aimed at strengthening education, including environmental and developmental matters, making it a universal right of peoples. This would provide the world with a flexible workforce capable of dealing with the growing environmental problems in development, ready not only to face change but also to influence it.

In forestry, education and training have to occur in institutions that are supported by the public sector, the private sector, or NGOs dealing with conservation and development. Trainers and professors should have a fairly good balance between practical skills and theoretical knowledge, but above all they should have a very good understanding of the major problems and opportunities existing in forestry development in their respective countries and regions.

Forestry education, then, should focus on a good balance between theory and practice, emphasizing technologies that are easy to adopt and easy to maintain. They must also have enough sensitivity to be able to absorb and utilize autochthonous technologies and knowledge of local origin, which in the end are cheaper and more effective.

Value of Local Knowledge

Thousands of farmers throughout the Third World have mixed trees and crops for many centuries in what today is called agroforestry. When CATIE in 1953 executed the first recorded research and coined the word *agroforestry* they put into perspective centuries' worth of trial-and-error efforts of numerous producers from the American tropics. Today agroforestry is a science of its own.

Long before formally trained scientists invented and popularized the term *agroforestry*, these farmers clearly recognized and explained the multiple uses and advantages of trees in polycultural farming systems: live fences, shade for coffee plants, forage, natural fertilizer from the tree litter, erosion control by the roots, and the nutritional and economic value of different indigenous tree species and their functions. Similar work has been done with multiple cropping patterns, pest control methods, soil fertilization and tilling, small animal husbandry, seed variety selection, uses of wild plant and animal species, unique botanical taxonomies, and curative herbs.

Forestry technicians and professionals must have motivation to work in the countryside. The challenge is to educate new professionals who can accept being away from the lures of large cities, with enough dedication and motivation to feel comfortable in and enlightened by rural communities, who can also adapt to working with private companies in the complexities of the sector.

A rural-people-to-rural-people training system can be instituted in which the groups of producers that have more experience or that have taken part in training workshops can teach their neighbors different techniques dealing with sustainable forestry. They also can transfer the best experiences with planting trees or making better use of the natural forests. This training system can impart a collection of methods and information related to conservation and land management that allows the rural families to resolve their own problems.

This approach is important, for it not only facilitates communication and dissemination of knowledge but also ensures that new practices and techniques draw on local knowledge and are tuned to local needs and capabilities. Moreover, through such an approach, the concept of participation is not restricted to simply ensuring the enthusiastic involvement of individuals and groups in a specific project or program. Participation is also about the capacity of local groups—particularly disadvantaged ones—to influence decision making and planning processes that affect their lives. Although development agencies throughout the region have clearly become more aware of the need for participation, their definition of participation is generally restricted to the former aspect.

It is crucial that basic technical knowledge be passed down to extension agents and local residents and that the technical packages be relatively simple. Here the question raised above—that of appropriate extension methodologies—is particularly important.

Research and Sustainable Forest Development

In order to achieve sustainable development, all countries need access to technologies and knowledge that would utilize all resources efficiently. Such technologies should be friendly to the environment and should cover scientific and technical knowledge in forestry as well as in economic, social, cultural, and environmental matters.

Obviously, both the generation and the transfer or adaptation of new science and technology are processes that would require a minimum local or national legal framework and infrastructure as well as the human resources that will be the key factors in advancing knowledge. Here there is a need for systematic training of laborers, technicians, managers, scientists, and educators. This group of people would become the cornerstone of the advancement of science and technology.

Governments should establish policy and legislation needed to encourage the generation of science and technology that are able to create a clear panorama of how the forest environment works, as well as a more precise idea of the capacity of the land to satisfy the demands of a growing population. Scientists can contribute to finding the optimal use of natural resources and energy in everyday activities in industry, agriculture and forestry, transportation, and other areas.

At the international level, science and technology generated at both regional and intercontinental centers—such as those in the Consultative Group for International Agricultural Research's Center for International Forestry Research (CIFOR), or the International Centre for Research in Agroforestry (ICRAF), or regional centers such as ASEAN forestry research centers in Asia or CATIE, which deals with forestry and agroforestry in Latin America—should be tapped. There is also the possibility of harnessing new forestry and related knowledge generated in the excellent national research centers located in various tropical countries in Latin America, Africa, and Asia.

New scientific knowledge can contribute to the sustainable management of the environment, assuring the everyday—and future—survival of humanity. Research can enable the human race to understand global or local phenomena related to climatic change, deforestation, pollution, and in general everything that affects the balance of nature. The new knowledge should then be used to design strategies for the achievement of sustainable development.

But we need to link sophisticated new knowledge with practical, local knowledge to find the best local solutions. That is why developed and developing countries require dedicated scientists to do research and to formulate recommendations. Developing countries need a critical mass of qualified scientists who can participate in a concerted effort to protect the environment and to achieve sustainability. It is essential that by the year 2000 the number of scientists in developing countries be increased several times over and that science and the generation of new knowledge be a common activity in most countries. Forestry should figure as a top-priority subject of research in countries with extensive forests as well as those with limited forests.

The U.S. National Research Council (1990) concluded that scientists must assume a leadership role in communicating their knowledge to policy makers; that research should incorporate an outreach component to communicate results of projects and activities to a broader range of clients whenever possible; that scientists should establish a professional reward system that acknowledges the validity of efforts of scientists involved in outreach; and that scientists must integrate extension specialists with their research counterparts in colleges, universities, and research centers in instances where interaction between the two groups is inadequate. The adoption of the new development paradigm will require forestry research to increase its breadth to include new areas and treat those areas in greater depth.

Fundamental means to carry out research and validation include personnel that are well trained, motivated, and with strong leadership; financial resources; physical resources including infrastructure, buildings, equipment, and field stations (even though in forestry the forest is the laboratory to do the research, recent advances in biotechnology need modern equipment); libraries; and sufficient scientific information. In addition, research needs to have suitable legal and political conditions if it is to flourish, leading to efficient organization and management of the work. The planners of such research should consider the long-term goals of the country in the terms of quality of life, exports, the protection of the environment, and the need to meet national demand for increased quantities of goods and services coming from the land. Some extranational considerations and global problems should also be taken into account in order to enhance the possibility of securing additional funding for forestry research from the international community. It is essential that the planning process reflect the priorities of the donor agencies and of international agreements.

Key elements in the planning process should be research actions aimed at understanding ecosystems (both functions and processes), understanding the relation between people and natural resources (focusing on social, cultural, environmental, and economic factors), and understanding the management and expansion of resource options (focusing on management practices, harvesting, utilization, trade, and conservation).

Genuine forestry development alleviates rural poverty and increases the production of goods and services. Forestry research must ultimately be judged against its contribution to this vision of development, and impact assessment must demonstrate how research leads to technology and how technology leads toward development goals or, alternatively, why it does not. Figure 6.3 shows different direct and indirect effects of forestry research.

The relation between agriculture and forestry has to be taken into consideration when planning and executing research. The general lack of understanding about the basic facts of natural resource systems and the degradation that can be caused by the intensity of some agricultural services can lead to specialization and stress in certain crop systems. Sustainable agriculture, in contrast, can promote the conservation of biodiversity, regenerate watersheds, slow down deforestation, and support many other ecological services, such as water conservation and wildlife preservation.

Figure 6.3 Direct and indirect effects of forest research

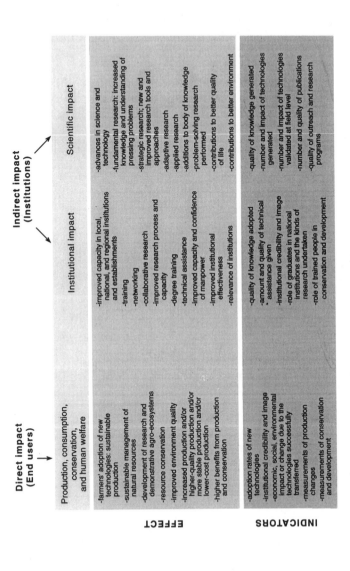

Direct impact
(End users)

Indirect impact
(Institutions)

	Production, consumption, conservation, and human welfare	Institutional impact	Scientific impact
EFFECT	-farmers' adoption of new technologies: sustainable production -sustainable management of natural resources -development of research and demonstrative agro-ecosystems -resource conservation -improved environment quality -increased production and/or higher-quality production and/or more stable production and/or lower-cost production -higher benefits from production and conservation	-improved capacity in local, national, and regional institutions and establishments -training -networking -collaborative research -improved research process and capacity -degree training -technical assistance -improved capacity and confidence of manpower -improved institutional effectiveness -relevance of institutions	-advances in science and technology -fundamental research: increased knowledge and understanding of pressing problems -strategic research; new and improved research tools and approaches -adaptive research -applied research -additions to body of knowledge -problem-solving research performed -contributions to better quality of life -contributions to better environment
INDICATORS	-adoption rates of new technologies -institutional credibility and image -economic, social, environmental impact or change due to the technologies successfully transferred -measurements of production changes -measurements of conservation and development	-quality of knowledge adopted -amount and quality of technical * assistance given -institutional credibility and image -role of graduates in national institutions and the kinds of research undertaken -role of trained people in conservation and development	-quality of knowledge generated -number and impact of technologies generated -number and impact of technologies validated at field level -number and quality of publications -quality of outreach and research programs

One of the challenges in forestry research has to do with marketing the idea that research is an investment, not an expense. Why do such countries as Japan, Germany, the Nordics, and the United States invest considerable amounts of money in this area? Why do smaller countries such as Costa Rica or emerging economies such as Chile invest in it, particularly in the private sector? They do so because they see forestry research as an investment that generates earnings. Selling this idea to a larger audience can be achieved by executing research that has been planned for the short, middle, and long terms, research that invests in future business opportunities and the general well-being of the population. In addition, research should be planned for high-priority areas where results will have a definite impact.

Developing countries cannot afford to dilute their scarce financial resources in trying to solve all their problems through research at once. They must establish priorities so that there exists a balance between basic and applied or adaptable research. There must also be a balance between research that is demand-driven and research that is supply-driven. When setting priorities they must take into consideration issues and factors that contribute to sustainability, so that it can be possible to include the forestry sector in sustainable development. And finally, research must be planned in such a way that the results are combined with efficient validation, transfer, and extension.

Most developing countries usually sacrifice more urgent needs, such as housing, health, basic education, or social security, to invest in research, and therefore they cannot afford research that is either irrelevant or not transferred opportunately. This is a very important policy question.

But investing in research is fundamental for development. New discoveries that have occurred through such investments have led humanity to surpass predictions of doom on several occasions, the most notorious being the Malthusian view that the world was heading toward collapse due to population growth and lack of food. More recently the Club of Rome, in "Limits to Growth" (1972), predicted that in the 1990s there would be chaotic conditions due to overpopulation and environmental problems. Development of science and technology through research has been responsible for overcoming such pessimistic forecasts.

Investments in research and development have to be planned well in advance if we are to reap the rewards of progress. Given the relatively long lag time between investment in forestry research and the resulting production increases and conservation effects, failure to invest today will show up in production or environmental degradation ten, twenty, or thirty years from now. The problems associated with environmental degradation will present themselves sooner. Not waking up until a global food or environmental crisis is upon society or until the last tree has fallen would be irresponsible.

Areas that have to be given a lot of thought in forestry research are those dealing with biotechnology and genetic engineering. There is a trend in developed countries where these endeavors are commonplace toward patenting and tying up intellectual property rights to all biotechnology and genetic engineering discoveries. It is also common to grant patents to species. In agriculture, patents issued to Agracetus/WRGrace for cotton and soybeans have given this company legally recognized rights and exclusive monopoly over any method of genetic transformation to any variety of these crops. This effectively discourages all other biotechnology research and development related to those crops, since researchers would be in violation of the patent if they did not pay a licensing fee or pay royalties to Agracetus.

The impact will be specially severe for breeders and molecular biologists in the public sector. Although the patents apply only in the countries where they are recognized, several aspects of the rapidly evolving International Trade and Patent System may give the claims global force. The recently adopted General Agreement of the World Trade Organization (WTO) requires that all countries adopt an intellectual property system for plants and microorganisms and allow signatory states to seek retribution from states that pirate patented technology.

Countries admitting species patents (the United States and countries covered by the European Patent Office, for example) could be within their rights to prohibit imports of raw materials or finished goods derived from a so-called pirated technology. Imports from India of clothing made from genetically engineered cotton or soybean paste from Brazil could be barred, for example.

The biodiversity convention that came into force at the close of 1993 is being interpreted in the United States and Europe as requiring all member states to acknowledge the patent rights of all member states when it comes to germ plasm collected prior to the adoption of the convention. On behalf of WRGrace the U.S. government would be within its rights to deny access to Grace's technologies or germ plasm to countries that do not recognize patent claims. Similarly, WRGrace would be within its rights to demand compensation from any country for the use of domestically developed genetic engineering on these crops even if the countries do not accept Grace's patents (Rural Advancement Foundation International 1994).

Outreach and Sustainable Forest Development

Outreach in forestry must involve suitable tools to train and motivate people and disseminate and transfer the research results, science, technology, and experience needed to convert the forestry subsector into the cornerstone of development, contributing significantly to the well-being of peoples. Outreach should also be aimed at strengthening institutions by convincing decision makers at high levels in government and in the private sector that forestry activities are profitable, that the resources used in research and education are not an expense but an investment, and that research can generate enough earnings to allow investment in more research, education, and outreach (figure 6.4).

At the technical level, forestry professionals should have adequate training in the management and utilization of forests and in the processing, marketing, and trading of for-

Figure 6.4 Strengthening forestry institutions and communities through outreach

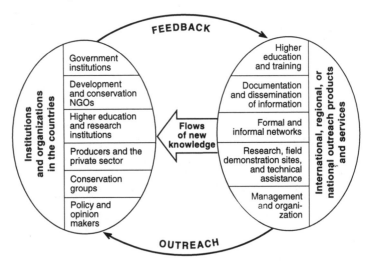

est products. They must have the capacity to work with people in rural communities and to understand their needs, and they must possess the tools to make local people participate in the solution of their problems.

Technology transfer groups should be aware of technologies available to farmers, producers, and forest dwellers that they should master in such a way that the technical assistance helps them build self-confidence. Feedback regarding transfer technologies should be provided to researchers so that necessary adjustments can be made to future technologies.

Nongovernmental organizations have proved to encompass many advantages in forestry development: continuity, presence at the local level, motivation, and understanding of local people. This is why outreach activities must take into consideration working with local and national NGOs in forestry extension, since they are fairly efficient in achieving long-lasting effects.

Private industry must also be brought to participate in the outreach activity because long-term feasibility of any forestry development depends on the value-added of industrial processing, marketing, and trading of forest and non-forest products. Strengthening the links at the outreach level between NGOs, industry, education and research institutions, and government would facilitate facing the challenge of finding or generating enough resources to strengthen their programs. There are many successful cases of such partnerships around the tropical world. Obviously, it takes people with entrepreneurial minds and experience in the private sector to identify products and services that could generate financial resources for the institution. In the case of research, there is the possibility of forming joint efforts with the private sector to finance specific applied research. Also, the organization of conferences and workshops should be aimed not only at training and knowledge dissemination but also as a good source of income through registration fees.

Linkages Between Education, Research, and Outreach

Interaction between education and training, research and validation, and outreach and technical cooperation must exist in order for sustainable forestry development to occur. The synergistic effect of all three activities working together, preferably under one umbrella, should focus their efforts at improving the standard of living and the well-being of people. Considerations from various fields dealing with social, biological, economical, environmental, agricultural, and cultural sciences should be taken into account (recall figure 6.1).

If any of these three important factors is missing, then these efforts are unlikely to achieve further stages of development and conservation. Universities, research centers, and development organizations have to find mechanisms to follow up the new knowledge as well as existing research and validation results, successful outreach, and technical cooperation programs. Many experts agree that this strategy has been the key to the success of agriculture in the United States, fueled by the work of land-grant universities that are managed under the concept of combining research, education, and extension.

According to CATIE (1994), the following are the criteria for prioritizing education, research, and outreach in forestry:

1. *Urgency.* There must be an urgent need and demand for a specific action.

2. *Economy of efforts.* There should be an effort to col-
laborate with local, national, international, and regional institutions and organizations, both public and
private, to minimize duplication of effort and increase
the effectiveness of actions.
3. *Institutional dimension.* There should be a rational
concentration of efforts according to the comparative
and competitive advantages of the institution, and its
historical strengths should be the ones emphasized in
the plan of action.
4. *Continuity.* The stability and durability of programs
and actions must be built, and their comparative and
competitive advantages continuously strengthened,
in order to be able to serve the beneficiaries in the long
term necessary for forestry research.

The absence of linkages between the generation and the
transfer of technology have resulted in completely untenable
situations in which the means used are irrelevant to the true
needs of farmers or transfer groups are unaware of the existence of certain relevant technologies. In the meantime, tech-
nology-hungry farmers strive to produce as much as possible
from their meager resources using outmoded means (Eponou
1993).

The following linkage-related problems are prevalent in
developing countries: missing steps in the transfer of technology, lack of mechanisms to join research and extension efforts, duplication of effort (both research and technology
transfer performing the same tasks), and inoperative or ineffective linkage mechanisms.

Some common mechanisms for linking research, education, and training, with full participation of local people, are
planning and review; execution of collaborative activities;
exchange of resources and knowledge; organization of conferences and workshops; dissemination of publications; eval-

uation, monitoring, and feedback; coordination of programs; and partnerships in the execution of various activities.

The challenge facing research institutions is to develop a greater capacity to facilitate effective interactions between researchers, technology transfer workers, and resource-poor farmers. This requires shifts in research policies and priorities, changes in the organization and management of research and technology transfer agencies, and the development of strong links between these agencies and the farmers. Here it is important that the researchers also have a direct link with farmers so that the research addresses directly the needs of those farmers and so that in their experimentation the researchers take local knowledge into account.

There are those who would oppose merging research and technology transfer on the basis that research could be jeopardized if it becomes involved in the development process. The luxury of separate research and technology transfer institutions can be enjoyed, however, only in developed countries with abundant resources for both. In most developing countries, the cost of doing research that is going to stay on the shelves of researcher's cabinets is too high. Therefore it is imperative that research and technology transfer be located under one umbrella.

Donors should be aware of the importance of these linkages. Donor involvement in financing education and training, research and validation, and outreach and technical cooperation in developing countries presently conforms to the following characteristics: lack of sustainability, discontinuity of effort, lack of coordination among agencies, and duplication of effort. Forestry research, education, and development should be seen as a unified long-term activity, covering at a minimum one rotation and preferably more. The complexities of field work today require the establishment of long-term research and demonstration plots, which will enable the gathering of information and experience on long horizons of time. This will in turn provide a stream of reliable information that will lead sooner rather than later to true and sustainable forestry development for the well-being of society in general.

These problems have been discussed in different fora at length, but up to now no one has taken the lead in solving them.

Summary

- Knowledge is not only the source of the highest-quality power but also the most important ingredient of force and wealth.
- Perhaps the most vital difference between developed and developing, rich and poor, is the gap in the capacity to generate, acquire, disseminate, and use scientific and technological knowledge.
- Empowerment through participation involves the recognition that local people are uniquely suited to decide what is best for their development. Thus, participation should be the basis for development.
- Training and education should include
  - strengthened attitudes and values;
  - formal and informal education;
  - socioeconomic and cultural aspects;
  - a balance between practical skills and theoretical knowledge;
  - understanding of the major problems and opportunities existing in forestry development; and
  - motivation to work in rural forested areas.
- Research needs
  - government policy and legislation to encourage science and technology generation;

- tapping of knowledge generated at regional and international centers;
- linkages between new knowledge and practical local knowledge;
- a critical mass of scientists;
- strong leadership;
- an outreach component to communicate results;
- understanding that research is an investment, not an expense;
- a focus on understanding ecosystems, people and cultures, economics, and management practices and options; and
- fiscal and financial resources.
- Outreach should
  - disseminate and transfer research results, science and technology, and experience;
  - strengthen institutions;
  - provide feedback to researchers; and
  - include NGOs, private industry, and entrepreneurs.
- Interaction between education and training, research and validation, and outreach and technical cooperation must take place in order for sustainable forestry development to occur.

CHAPTER SEVEN

The Economics of Deforestation

DAVID PEARCE

As discussed in Chapter 5, high rates of deforestation have persisted despite considerable attention to national policies and programs. Loss of all forest is very unlikely since topography, climate, and distance from markets will result in a remaining core of forests. But the value of that remaining core will be severely reduced as a source of biological diversity and economic activity. Loss of "critical minimum size" and the absence of linking corridors of forest will, among other factors, limit the functions of the remaining forest.

It is usual to place the blame for deforestation on various factors: the greed of logging companies and large livestock ranchers, poverty itself as the poor are obliged to "mine" forest resources in the absence of alternative sustainable livelihoods, unthinking governments, the side effects of road building (which opens up forest areas), the sheer pressure of population growth, and so on. All of these are relevant, but it seems fair to say that much of the conventional wisdom on deforestation and its causes stems from a failure to analyze the underlying causes of deforestation. Rather than pointing to the slash-and-burn cultivator as causal agent, what is needed is an investigation into why farmers behave in the way they do. Much the same goes for logging companies: Why do they exploit at the rate they do rather than at some sustainable rate?

It is only when we trace the causal analysis through to the workings of the economic system that fundamental forces can be identified, and only if these forces are identified can we begin to determine the policy response. For past policy has all too often produced solutions without causal analysis. The result has been protected areas in which protection is more apparent than real because the original structure of incentives giving rise to deforestation has remained unaltered. If, for example, the original problem lies in overuse of the forest by surrounding populations, simply demarcating the area as "protected" does little to alter the behavior of those who are now excluded from the area. The more recent focus on developing and demonstrating sustainable forest management (SFM) systems is to be welcomed since this at least focuses attention on the need to provide acceptable alternatives for forest users. But even in demonstrating SFM systems there has been little analysis of the comparative rates of financial return to sustainable and unsustainable alternatives. Basically put, unless sustainable management yields higher returns, there will be an incentive to continue unsustainable practices.

This chapter sets out a very simple model for comparing sustainable and unsustainable systems in the forest context. Its purpose is to point to the factors that should be investigated before interventions are designed.

The Nutrient Mining Model

The model investigated here is based on the idea of "nutrient mining" (Schneider 1992, 1994). *Nutrient mining* refers to a process whereby the biomass of the tropical forest is converted to a stock of nutrients that are then treated as an exhaustible resource to be exploited. The conversion process involves burning the biomass so that nutrients stored in the biomass (with the exception of nitrogen, which is utilized in complete combustion), usually by far the greatest source of nutrients in a tropical forest, are stored in the resulting ash. The ash is then the resource on which the subsequent land use relies. But it is a finite quantity. The fire also serves as a pesticide, clearing the land of pests so that crop growing or cattle ranching can take place. Once exhausted of the released nutrients (or, as we shall see, well before then), the land is abandoned for new "frontier" land. The previous land may then revive as secondary forest if nutrients have not been totally lost, remain degraded and ecologically unproductive if severe nutrient depletion has taken place, or be colonized by some use that can adapt to lower nutrient value. New land is subjected to the same process. Since "old" land is not (generally) reused by the colonizers, the process is one of systematic deforestation.

This model of nutrient mining assumes that there is a frontier—that there is always new forestland to be converted to agriculture. As such, it fits best the context of South America, where, in general, new land is available. Where the frontier is "closed" (as in Asia; see Chapter 3), and there is no ready supply of new land, deforestation is a more dynamic process. It could, for example, be characterized by slash-and-burn agriculture, in which land is abandoned well before total nutrient loss, allowed to remain fallow, and used again at a later stage. Here the picture is complicated by rapid population growth that makes an ostensibly sustainable system unsustainable. Fallow periods get shorter and shorter and land is degraded. In some cases the absence of frontier land actually results in beneficial conservation policies. Realizing that there is nowhere to go, farmers settle on the land they have got and introduce conservation practices. This appears very much to have been the experience in Machakos, Kenya, where previously degraded land has been restored and is farmed profitably (English, Tiffen, and Mortimore 1994). Most of the features of the nutrient mining model can be transposed to the slash-and-burn context.

The Colonizing Decision: A Diagrammatic Treatment

Figure 7.1 shows the nutrient mining model in simple outline. On the vertical axis is money (shown here as dollars), and on the horizontal axis is time. The declining curve A shows the annual profit to be obtained from agriculture or ranching based on the mining of the nutrients once the land is cleared by burning. Although the decline in profitability need not be as smooth as shown in the diagram, evidence from a limited number of studies suggests that it is as shown (Schneider 1994). The second declining curve, B, shows the

Figure 7.1 Nutrient mining model

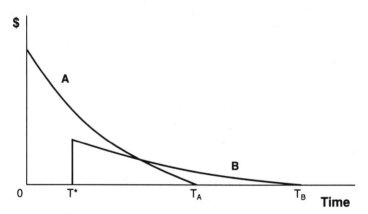

same situation but for a further area of forest. This curve has a lower initial height because it is farther away from the market than the first plot of land, thus raising transport costs and reducing profit. Or, land quality may decline as the extensive margin is colonized. Both plots are accessible, but the first has higher profitability and hence is chosen first. If plot A is farmed to the point of nutrient exhaustion, it will be farmed up to point TA. Then the second plot will be cleared and farmed until point TB is reached, and so on. In practice the colonizing decision is more complex than this and the decision to move on to plot B will be taken before time TA. Essentially, although the second plot has lower profitability than the first, there will be some point in time, such as T*, when it will pay to move to the second plot and abandon plot A. This will occur when the farmer perceives that the future expected profits from B exceed the profits remaining from the exploitation of A. Thus, although complete nutrient exhaustion might take ten to fifteen years, in practice the plot will be farmed for only five to seven years, a typical "residence time" in frontier agriculture. What happens to plot A between T* and TA depends very much on alternative uses. The initial colonizer could be a logger, so that the residence time is very short. The forest roads created by the logging company open up the forest and permit agriculturists to move in. The first user may grow crops and abandon the plot at T*, and ranching may succeed that until the rancher also perceives that it is profitable to switch plots. The simplicity of figure 7.1 thus disguises what may well be a complex situation.

The presence of time in figure 7.1 suggests that its effect on farmers' decisions needs to be taken account of explicitly. Nothing is lost if figure 7.1 is interpreted as showing what the economist calls present values of expected profits. A present value is the value of future profits expressed in units of currency now. Effectively, future profits are discounted not because of any issues relating to inflation but because of the colonizer's preference for receiving his returns now rather than later. Simple impatience could explain the presence of discounting, but there is a force at the frontier that is likely to be particularly powerful and that will make for high discount rates (that is, for a low value being placed on the future as compared to the present). This force is uncertainty. The more uncertain the future, the less incentive the individual has to invest in conservation of the land. Hence the kind of positive feedback from land scarcity to conservation, as in the Machakos case, will be missing if the future is uncertain and if there is new land for the taking. Surprisingly, evidence on what farmers' discount rates actually are is extremely limited. The only detailed and rigorous study is that of Cuesta, Carlson, and Lutz (1994), which suggests that farmers in Costa Rica may be discounting the future at such a high rate that even conventional soil conservation practices are not profitable when viewed from the farmer's standpoint.

Why is uncertainty so powerful a force in this case? The dominant factor generating uncertainty will be lack of resource and land tenure. If rights to the land are poorly defined, or, as in many cases, not defined at all, then the occupier of the land cannot be sure of continued occupancy. It will always be under threat from some other potential owner. Even if land rights are fairly well defined, rights to the products of the land may be attenuated, meaning that it is always important to assess both land rights and usufructuary rights. Absence of land tenure makes investment in the existing plot of land even less likely, since the absence of tenure also tends to make the securing of credit difficult or expensive. If credit is rationed or expensive, the effect is to raise the colonizer's discount rate further, since future profits have now to cover high rates of interest.

Sustainable Forest Use

Figure 7.1 encapsulates the basic process of nutrient mining. But it is still only half the picture, since there is usually some alternative way of managing the land that is sustainable. This may be based on other forest products such as nuts, latex, wild meat, or agroforestry crops. Much has been made of the rates of return that can be expected from such uses, and some studies indicate that sustainable management has higher rates of return than, for example, clearance for timber (Peters, Gentry, and Mendelsohn 1989). Although the evidence is still limited, what there is suggests that there is not a strong case for the conservation of tropical forests on grounds of local economic value alone. The initial enthusiasm that greeted claims of high-value sustainable forest use based on nontimber forest products (NTFPs) has waned somewhat in the face of (a) methodological doubts about such studies (for example, Godoy and Lubowski 1992; Southgate and Whitaker 1994), (b) revised estimates of NTFP productivity, (c) wrong extrapolation from one forest type to another (Phillips 1994; Godoy and Lubowski 1992), and (d) doubts about the sustainability of NTFP exploitation itself (Peters 1991). For these reasons, sustainable use profitability initially would lie below the profitability of unsustainable use shown in figure 7.1. This is not a necessary result: there is certainly evidence that sustainable forest use practices are more profitable in certain contexts (Pearce and Moran 1994). But if widespread opportunities for profitable, sustainable practice exist, it is legitimate to ask why they have not been adopted. Undoubtedly, lack of information, poor or nonexistent extension services, and traditional inertia play their roles. But the overall profitability of sustainable use when viewed from the standpoint of the colonizer is questionable.

The picture in figure 7.1 is now complete. Nonsustainable use of forestland is seen to be at least initially more profitable for the farmer than sustainable use, and it could well be more profitable over the period any farmer may plan for. If the picture is as suggested in figure 7.1, then the problems of finding sustainable uses for forestland that are consistent with sustainable livelihoods for forestland users and with conservation are formidable. This suggests a closer look at the factors determining the positions of the various curves.

Discounting

The relevance of the discount rate was noted earlier. In fact the discount rate may well be the most important factor giving rise to nonsustainable management of forestland. Of course, poverty itself tends to be associated with high discount rates. Anyone on the margin of poverty is unlikely to look very far into the future if the concern is with food today and tomorrow rather than in several years' time. But the exact role that poverty plays in accelerating deforestation is far from clear. First, if deforestation is associated with poverty one would expect this to show up in statistical studies of the factors giving rise to deforestation. Most of the available studies are collected together in Brown and Pearce (1994). Table 7.1 summarizes the results. They suggest that there is no absolutely conclusive link between any of the variables analyzed and deforestation. But cautious conclusions might include the following:

1. The balance of evidence favors the hypothesis that population growth has an impact on deforestation;

2. Population density is clearly linked to deforestation rates;
3. Income growth is fairly clearly linked to rates of deforestation, suggesting that deforestation has more to do with growth of incomes than with poverty—a result that runs counter to the popular interpretations of the causes of environmental degradation;
4. The evidence on the role of agricultural productivity change is finely balanced. One would expect growth in productivity to lessen the pressure on colonization of forests, that is, the coefficient of association should be negative. The two studies finding this association are for South America and Indonesia. The two studies finding the opposite association are for Thailand and the Brazilian Amazon; and
5. The link between country indebtedness and deforestation is ambivalent. One study finds a positive link for tropical moist forests but not for other forests; another finds a similar positive link, and another finds no such link. Again, this ambivalence is at odds with the popular interpretations for the causes of environmental degradation.

Surprisingly, few of the studies available at the time of this survey accounted for property rights regimes, despite the fact that property rights are cited as a major factor in environmental degradation generally and in the theory of tropical forest loss in particular. The exception in table 7.1 is the work of Southgate, in which property rights in South America are seen to be an important explanatory factor. More recent work by Deacon (1994) underlines the importance of property rights. Deacon further relates these rights, or the lack of them, to government instability and lack of democratic participation. Although the evidence in table 7.1 is interesting, it is of limited value, since the processes of deforestation are complex and not easily captured in statistical

Table 7.1 Econometric Studies of Deforestation

Deforestation significantly related to:	Rate of population growth	Population density	Income	Agricultural productivity	International indebtedness
Allen and Barnes 1985	+				
Burgess 1992		+	−		+
Burgess 1991 a)	−		+		+
Burgess 1991 b)	−				
Capistrano and Kiker 1990 a)			+		−
b)			+		−
c)	+		+		
Constantino and Ingram 1991	−	+	−	−	
Kahn and McDonald 1994				+	+
Katila 1992		+			
Kummer and Sham 1991		+			
Lugo, Schmidt, and Braun 1981		+			

Study					
Palo, Mery, and Salmi 1987			+		
Panayotou and Sungsuwan 1994			+	−	
Perrings 1992 a)	+		+	+	
Perrings 1992 b)	−		+	+	+
Reis and Guzman 1992	+	+			
Rudel 1989	+	+			
Shafik 1994					
Southgate 1994	+				
Southgate, Sierra, and Brown 1989	+	−			

Source: Brown and Pearce 1994.

A minus sign (−) means that an increase in the variable leads to a decrease in deforestation.

A plus sign (+) means that an increase leads to an increase in deforestation.

Blank entries mean either not statistically significant or not tested for.

Additional references: Brown, K. and Pearce, D. W. 1994. *The Causes of Tropical Deforestation: The Economic and Statistical Analysis of Factors Giving Rise to the Loss of the Tropical Forests.* London: University College London Press.

Burgess, J. 1991. Economic analysis of the causes of tropical deforestation, M.Sc. Thesis, Department of Economics, University College London.

Capistrano, D., and C. Kiker. 1990. Global economic influences on tropical closed broadleaved forest depletion, 1967–85. Paper to the International Society for Ecological Economics Conference, Washington D.C.

Lugo, A. E., R. Schmidt, and S. Brown. 1981. Tropical forests in the Caribbean, *Ambio*, 10, 6, 318–24.

Shafik, N. 1994. Economic development and environmental quality: An econometric analysis. *Oxford Economic Papers*, 46, 757–773.

associations, and notably so when the data that researchers
are obliged to use are so unreliable. Moreover, none of the
studies surveyed included any explicit or surrogate variable
for the discount rate. Yet Schneider (1994) shows that the discount rate in the Amazon context is of major significance.
Sustainable uses would have to have profits of around 80 percent of those of unsustainable uses in the initial year if sustainable use is to be preferred to unsustainable uses, so powerful is the role of the discount rate.

But this finding points the way toward policy conclusions. First, if poverty and high discount rates are corre-
lated, then traditional development programs aimed at raising per capita incomes are likely to be environmentally
beneficial through the effect of lowering the discount rate.
This finding is somewhat uncertain, however, because of
the very sparse evidence available on the link between incomes and discount rates. A more targeted conclusion
would be that policies aimed at improving credit facilities
for colonizers could result in the removal of one of the constraints on land improvement, which in turn should reduce
the incentive to colonize the frontier. Put another way,
credit facilities could enable increased profitability on the
first exploited piece of land (plot A in figure 7.1), reducing
the relative attractiveness of frontier plots (plot B). Of
course, if cheaper credit is made generally available, it may
do little to prevent frontier expansion: the cheaper credit is
as available for plot B as it is for plot A. But if credit can be
targeted at initial land areas and denied for frontier land,
then the incentive to remain on the existing land and improve its productivity would be stronger. This suggests a
differential credit policy as one of the means of encouraging land retention. Note that even this policy does not result

in a switch from unsustainable to sustainable use, but it does slow the rate of deforestation.

Agricultural Productivity

In the same vein, extension services and even the provision of agricultural inputs could help raise the productivity of existing land, reducing the incentive to extend into the frontier. Once again, the effect is likely to be one of slowing the rate of deforestation but without altering the choice of land management technique; that is there is no incentive here to switch to the sustainable alternative (figure 7.2).

Figure 7.2 Worker extracting resin from a pine tree

Tenure

If rights to the land can be conferred, then, as previously discussed, there will be more incentives to stay on the land than to make a frontier conversion.

Removing Economic Distortions

Figure 7.1 assumes that sustainable and unsustainable uses are being considered on an equal footing. But in practice it is well known that there are substantial economic incentives to deforestation through the provision of subsidies for land clearance and for logging. The examples are, by now, well known (Pearce and Warford 1993) and include the subsidies to forest conversion for livestock in Brazil up to the end of the 1980s (although their importance is now disputed—see Schneider 1994); the failure to tax logging companies sufficiently, giving them an incentive to expand their activities even further; and the encouragement of inefficient domestic wood processing industries, effectively raising the ratio of logs, and hence deforestation, to wood product. What intervention does is distort the competitive playing field. Governments effectively subsidize the rate of return to land conversion, tilting the economic balance against conservation. In terms of figure 7.1, the effect is to shift the profitability curve of the unsustainable use outward to the right, making it more attractive still than the sustainable use.

Table 7.2 assembles some information on the scale of the distortions that governments introduce. Such distortions are widespread. The general rule in developing countries is for agriculture to be taxed, not subsidized, but significant subsidies exist in several major developing countries, such as Brazil and Mexico. By comparison, OECD countries are ac-

Table 7.2 Economic Distortions

A. Agricultural producer subsidies[1]

Mexico	mid-1980s	+53%
Brazil	mid-1980s	+10
South Korea	mid-1980s	+55
Sub-Saharan Africa	mid-1980s	+09
OECD	1992	+44

B. Timber stumpage fees as percentage of replacement costs[2]

Ethiopia	late 1980s	+23
Kenya	late 1980s	+14
Ivory Coast	late 1980s	+03
Sudan	late 1980s	+04
Senegal	late 1980s	+02
Niger	late 1980s	+01

C. Timber charges as percentage of total rents

Indonesia	early 1980s	+33
Philippines	early 1980s	+11

Sources: Agricultural PSEs: Moreddu et al., 1990; OECD 1993; stumpage fees; World Bank 1992; and Repetto and Gillis 1988.

[1]Producer Subsidies are measured by the "Producer Subsidy Equivalent" (PSE) which is defined as the value of all transfers to the agricultural sector in the form of price support to farmers, any direct payments to farmers, and any reductions in agricultural input costs through subsidies. These payments are shown here as a percentage of the total value of agricultural production valued at domestic prices.

[2]A stumpage fee is the rate charged to logging companies for standing timber. It is expressed here as a percentage of the cost of reforesting (b) and as a percentage of total rents (c).

tually worse at subsidizing agriculture. In 1992, OECD subsidies exceeded $180 billion (OECD 1993). These subsidies work in two ways. Subsidies in developing countries will tend to encourage extending agriculture into forested areas. Subsidies in the developed world make it impossible for the developing world to compete properly in international markets, often locking them into subsistence agricultural practices. Although the removal of OECD country subsidies would appear to be a recipe for expanding land conversion in the developing world in order to capture the larger market, the demands of a rich overseas market are more likely to result in agricultural intensification and hence reduced pressure on forested land.

Table 7.2 also shows that many developing countries fail to tax logging companies adequately, thus generating larger "rents" for loggers. The larger rents have two effects: they attract more loggers and they encourage existing loggers to expand their concessions and, indeed, to do both by persuading the host countries to give them concessions. Persuasion involves the whole menu of usual mechanisms, including corruption.

Distortions do not only operate through the agricultural and forest sector. Road-building plays a major role in opening up forest areas. Yet the social viability of many roads is often not tested, and their construction may often have more to do with "grand design" and military objectives.

Global Appropriation Failure

Market failure describes the inability of existing markets to capture nonmarket and other economic values of the forest within the context of the country or local area. But there are missing global markets as well. These failures are very relevant to the apparent superiority of nonsustainable forest uses. We can consider two such global markets that are highly relevant to tropical forests: the nonuse, or existence, value possessed by individuals in one country for wildlife and habitat in other countries, and the carbon storage values of tropical forests. Global appropriation failure arises because these values are not easily captured or appropriated by the countries in possession of tropical forests.

Nonuse Values

Economists use methods of measuring individual preferences as revealed through individuals' willingness to pay
(WTP) to conserve biodiversity. "Global valuations" of this
kind are still few and far between. Pearce (1995) assembles
the results of contingent valuation methods (CVMs) in several countries. These report willingness to pay for species
and habitat conservation in the respondents' own country.
They suggest that households may well be willing to pay
around $10 per annum for broad forest and species conser-
vation. These studies remain controversial. In the context of
tropical forests and biodiversity this controversy has some
justification. In particular, "embedding"—the problem of
valuing a specific asset rather than the general context of
which the asset is part—is bound to be a major problem for
assets that are remote from respondents or jointly produced
with other assets (for example, species within prized habitats)—see Schulze, McClelland, and Lazo (1994). Although
we cannot say that similar kinds of expressed values will
arise for protection of biodiversity in other countries, even a
benchmark figure of $10 per annum for the rich countries of
Europe and North America would produce a fund of $4 bil-
lion per annum, roughly four times the size of the fund that
is available to the Global Environment Facility in its operational phase and perhaps ten times what the fund will have
available for helping with biodiversity conservation. Clearly,
a focal point for biodiversity conservation must be the conservation of tropical forests.

It is also possible to look at the implicit prices for conservation in debt-for-nature swaps. The procedure of estimating
implicit prices of this kind is open to doubt, although it has
been used by some writers—see Ruitenbeek (1992) and

Pearce and Moran (1994). Numerous debt-for-nature swaps have been agreed. It is not possible to be precise with respect to the implicit prices since the swaps tend to cover not just protected areas but education and training as well. Moreover, each hectare of land does not secure the same degree of protection, and the same area may be covered by different swaps. In addition, Pearce and Moran (1994) arbitrarily chose a ten-year horizon in order to compute present values, whereas the swaps in practice have variable levels of annual commitment. The range of implicit values is from around one cent per hectare to just over four dollars per hectare. This is small compared to the opportunity costs of protected land, although if these implicit prices mean anything they are capturing only part of the rich world's existence values for these assets. That is, the values reflect only part of the total economic value.

Finding a benchmark from such an analysis is hazardous, but something on the order of $5 per hectare may be appropriate. If so, these implicit existence values will not save the tropical forests. On the other hand, debt-for-nature swaps clearly involve many free riders since the good in question is a pure public good and the payment mechanism is confined to a limited group. Looked at another way, $5 per hectare per annum for saving, say, 25 percent of the world's remaining closed tropical forests would amount to a fund of roughly $1 billion per annum (780 m ha × $5 × 0.25).

The only intercountry valuation exercise appears to be that of Kramer, Mercer, and Sharma (1994). This reports average WTP of U.S. citizens for protection of an additional 5 percent of the world's tropical forests. One-time payments amounted to $29–$51 per household, or $2.6–4.6 billion. If this WTP were extended to all OECD households, ignoring income differences, a broad order of magnitude would be a one-off payment of $11 to 23 billion. Annuitized, this would be, say, $1.1 to 2.3 billion per annum.

All these "global" estimates are very crude, heroic even, but it is interesting to note that the hypothetical payments are not wildly divergent:

	$ billion per ann.
Implied WTP (GEF)	$0.4
Implied WTP (DFN)	$1.0
Like Assets Approach	$4.0
Global CVM	$1.1–2.3

Carbon Storage

All forests store carbon, so that, if cleared for agriculture, there will be a net release of carbon dioxide, which will enhance the greenhouse effect and hence potentially speed global warming. In order to derive a value for the "carbon credit" that should be ascribed to a tropical forest, we need to know (a) the net carbon released when forests are converted to other uses and (b) the economic value of one ton of carbon released to the atmosphere. Data collected in Brown and Pearce (1994) suggest that, allowing for the carbon fixed by subsequent land uses, carbon released from deforestation of secondary and primary tropical forest is on the order of one hundred to two hundred tons per hectare. Evidence from Fankhauser (1995) suggests that each ton of carbon does damage valued at at least $20. Using these figures we can conclude that converting a forest to agriculture or pasture would result in global warming damage of, say, $2,000–$4,000 per hectare.

How do these estimates relate to the development benefits of land use conversion? We can illustrate with respect to the Amazon region of Brazil. Schneider (1992) reports upper-bound values of $300 per hectare for land in Rondonia. The figures suggest carbon credit values two to fifteen times the price of land in Rondonia. These carbon credits also compare favorably with the value of forestland for timber in, say, Indonesia, where estimates are on the order of $1,000–$2,000 per hectare. All this suggests the scope for a global bargain. The land is worth $300 per hectare to the forest colonist but several times this to the world at large. If the North can transfer a sum of money greater than $300 but less than the damage cost from global warming, there are mutual gains to be obtained.

Pearce (1995) addresses the issue of such global bargains. From a policy standpoint it means that field personnel should always be sensitive to the possibility that not only may a given forest area be suitable for GEF grants, but that there may be private sector openings, too.

Expanding Sustainable Management Profitability

Figure 7.1 also makes it clear that if sustainable uses can be made more profitable, then there will be an incentive to switch to them. This is certainly an agenda for the future. Some "biodiversity prospecting" is already taking place. These contracts involve the exchange of plant genetic material for funds that are paid directly to the local community, as with the various pharmaceutical deals between Biotics Ltd., Merck Pharmaceutical, and Pro Natura's canopy balloon. What is important here is that finding additional products and services is likely to be more profitable than finding a complete alternative to available sustainable uses of forestland.

Conclusions and Summary

Securing the sustainable management of tropical forests is a complex task. It will not be achieved by simply declaring an area to be protected, valuable as protected areas are in securing conservation objectives. What is needed is a closer look at the underlying structure of incentives for forest destruction. This suggests a focus on the following areas for maximum exploitation of sustainability:

- Prevailing distortions: subsidies, concessions, tax regimes, infrastructure construction
- Land tenure and resource rights to lower discount rates
- Differential credit availability and price, also affecting the discount rate
- Productivity-increasing ventures on already converted land
- "Global bargains," either through the GEF or through the attraction of private-sector deals involving biodiversity prospecting and carbon offsets.

The end result is a package of measures generated by what some call a new economics but that in reality is an application of well-tried economic theories.

CHAPTER EIGHT

Creation of Country-Specific Strategies

JOHN SPEARS

Earlier chapters of this book have mentioned the importance of developing stronger cross-sectoral linkages between forestry and other sectors and the need for a balanced approach to formulation of forestry strategies that, in addition to emphasizing the environmental benefits of forest conservation, recognizes the potential of forests to contribute to poverty alleviation, sustainable agriculture, and economic growth. Also, the book has stressed the need for radically reformed approaches to development of national forest policies to ensure the full participation of local people and NGOs in policy dialogue.

This chapter is concerned with the implications of these emerging approaches to the formulation of country-specific, integrated forest strategies and for national capacity-building.

The Rio Earth Summit Principles

As a starting point, it is useful to refer to the Rio Earth Summit Forest Principles, which were the beginning of a global consensus on forests. Despite the strong emphasis during the Rio forestry debate on the sovereign rights of individual countries to manage forest resources for maximization of national benefit, it was significant that all countries demonstrated a willingness to recognize demands that do not necessarily coincide fully with national priorities.

The discussions in this chapter take into account the desirability of factoring into national policies recognition of the role that national forests may also play in contributing to international trade and to regional and global environmental protection.

This chapter places special emphasis on the need for alternative, more broadly based approaches to national forest strategy formulation that are devised with consideration for local people.

Toward Sustainable Conservation and Development of National Forest Resources

A recurrent theme of the post-UNCED debate on forest conservation and management objectives has been the desirability of reconciling the many different perspectives on what constitutes "sustainable" forest management. Different stakeholders have very different perceptions on issues such as the relative merits of multiple versus single end uses, definition of the geographic boundaries of sustainability (the individual forest? the national forest estate? the entire landscape?), and the timetable for achieving sustainability.

Many alternative definitions of sustainable forest management have been proposed. One that seems effectively to capture the concerns of most of these alternative formulations and that has been adopted as a basis for the ensuing discussions in this chapter is the following: sustainable forest management balances the production of forest goods and services to meet varied human needs on a continued basis and in ways that leave the various ecosystem capacities of forestland unimpaired.

Sustainable management is a continual process rather than a definite end point, and precise definitions of sustainable management need to be locally negotiated and adjusted to reflect varying cultural, socioeconomic, and ecological circumstances. Some of the more important factors that will influence the transition toward sustainable forest management are summarized below.

Key Elements of an Integrated National Forest Strategy

As an overriding principle, a participatory approach to national policy formulation with a built-in provision for continuous improvement is desirable. Within that framework, some of the key elements of the process needed to develop an integrated forest strategy are as follows:

- Participation of all stakeholders
- Environmental objectives
- Economic objectives
- Social objectives
- Intergenerational concerns
- Integration of the various objectives
- Economic, environmental trade-offs
- Precautionary approaches
- Ecologically sound practices
- Provision for monitoring and evaluation
- Provision for continuous improvement and revision

*Ensuring Stakeholder Involvement in Debate,
Decisions, and Management*

Figure 8.1 suggests a conceptual approach for developing an integrated national forest strategy that recognizes local stakeholders' perceptions as the starting point for setting national sustainable forest management goals. Government at all levels, indigenous people's and women's groups,

Figure 8.1   Integrated national strategy

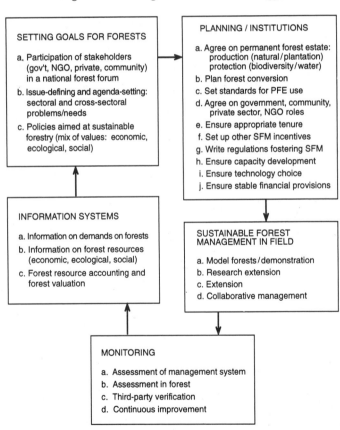

SETTING GOALS FOR FORESTS

a. Participation of stakeholders
   (gov't, NGO, private, community)
   in a national forest forum
b. Issue-defining and agenda-setting:
   sectoral and cross-sectoral
   problems/needs
c. Policies aimed at sustainable
   forestry (mix of values: economic,
   ecological, social)

PLANNING / INSTITUTIONS

a. Agree on permanent forest estate:
   production (natural/plantation)
   protection (biodiversity/water)
b. Plan forest conversion
c. Set standards for PFE use
d. Agree on government, community,
   private sector, NGO roles
e. Ensure appropriate tenure
f. Set up other SFM incentives
g. Write regulations fostering SFM
h. Ensure capacity development
i. Ensure technology choice
j. Ensure stable financial provisions

INFORMATION SYSTEMS

a. Information on demands on forests
b. Information on forest resources
   (economic, ecological, social)
c. Forest resource accounting and
   forest valuation

SUSTAINABLE FOREST
MANAGEMENT IN FIELD

a. Model forests/demonstration
b. Research extension
c. Extension
d. Collaborative management

MONITORING

a. Assessment of management system
b. Assessment in forest
c. Third-party verification
d. Continuous improvement

NGOs, universities, research institutions, and the private
sector all have a role to play in that process. Inevitably, in
the course of consultation between these various stakeholder groups there will emerge the need for national-, re-
gional-, and global-level conflict resolution and consensus-
building mechanisms. That issue is further discussed
below.

Defining and Prioritizing the Major Issues

Reaching agreement on the key issues (both sectoral and
cross-sectoral) and on the broad forest agenda requires considerable stakeholder interaction and should be a reitera-
tive process. A common weakness of many past forestry
sectoral planning exercises has been their failure to focus
on key elements of policy and institutional change and
on the essential linkages between forestry and other sectors.
Both are fundamental to definition of effective national
programs and to reducing external pressures on forest resources.

Prioritization in the setting of national goals is a key issue. When setting priorities, in addition to more traditional
concerns such as timber production, revenue generation,
and protection of watersheds and wildlife, the important role
that forests can play in poverty alleviation merits special
consideration. There are now available many analyses of the
linkages between sustainable use of forest resources and family nutrition, health, food security, sustainable agriculture,
and rural income generation. Historically, national policy
statements have largely ignored or understated the potential
of forests to contribute to poverty alleviation and the central
role that forests play in meeting the basic needs of the rural
poor.

Policy and Institutional Reforms

To accelerate the current transition throughout both developed and developing countries from centralized forest-sector-oriented management to more broadly based participatory and ecologically based forest conservation and development will frequently require reform of existing regulatory, fiscal, and institutional policies. Early identification of the areas of policy reform that are most essential for effective forest conservation and development is one of the most important elements of national policy formulation.

One key issue will be that of assuring stability of tenure and preservation of local communities' rights of access to forest resources. Regulatory and fiscal incentives can also play an important role by creating a framework conducive to local-community and private-sector involvement in the sustainable harvesting, consumption, and management of forest resources.

A problem area that has proved difficult to address in the past has been the inability of weak government forestry institutions to persuade central government economic planning and similar bodies to tackle what are frequently politically sensitive reforms. Policy and institutional reforms that influence sustainable management of forest resources and more equitable distribution of forest benefits include

Regulatory instruments

- Land-tenure arrangements and land-use zoning
- Legislation to protect traditional property rights and access to forestlands
- Legislation to protect intellectual property rights to germ plasm
- Trade regulations

- Timber concessions, license agreements, and penalties for noncompliance
- Legislation to protect biodiversity reserves and national parks
- Global conventions with implications for forest-resource management

Financial instruments

- Removal of taxes and subsidies that encourage deforestation
- Revenue-sharing contracts with local communities
- Stumpage fee and timber harvesting or export tax levels that capture economic rent

Planning a Forest Strategy

Historical approaches to forest land-use planning were frequently based on a conviction that a major responsibility of state forestry agencies was to protect and manage a government-owned forest estate on behalf of local communities and for national environmental benefit. A key element of this strategy provided for the exclusion of people and livestock from government-controlled forestlands.

The flaws in this territorially driven forest conservation strategy have now been widely recognized. There are many countries in which government forest agencies remain reluctant to devolve responsibility of state-owned forestlands to local communities, however. Earlier forest reservation policies were often implemented in times when local population pressures and rural poverty were less serious problems. Forest "reservation" processes were primarily concerned with creating and protecting an adequate forest "estate."

One consequence of this policy was that substantial parts

of what were reserved in earlier days as government forests were located on relatively flat lands overlying soils with reasonable agricultural potential. Given today's growing population pressure on many tropical forested areas and widespread rural poverty, further encroachment of forestlands is inevitable. There are three distinct phases in planning a forest strategy.

Phase 1

The first step is identification of strategic areas that should be retained as the core of a permanent forest estate. These will include biodiversity reserves, vital water catchments, national parks, wilderness areas, and wood-production forests containing valuable slower-growing species, management of which implies very long rotations. Even though these areas will remain under government control, recent history has clearly demonstrated that survival of these "reserves" will largely depend on the extent to which the integrated forest strategy succeeds in fully involving local communities. This in turn will depend on the extent to which local communities benefit from the management decisions taken (figure 8.2).

Phase 2

The next step is identification of areas of reserved forest that could safely be converted to more intensive land use without the risk of serious ecological damage. A key issue is the need for a more flexible attitude on the part of traditional forest administrations to "degazettement" of carefully selected forest areas.

There is well-documented evidence that local communities and farmers are highly aware of the value of the products and services provided by forests, trees, and open woodlands. In the course of conversion to agriculture it is quite common for them to protect more valuable indigenous tree species and deliberately to retain patches of remnant woodland as grazing reserves, as a source of products (thatching grass, forest fruits, medicinal and other products), or both.

The net result is replacement of closed natural forests by agroforestry farming systems that frequently combine retention of selected indigenous trees with newly planted and often faster-growing fruit and multipurpose trees.

Recognition of the desirability of this transition in national forest policy statements would remove one potential source of tension between forest administrations and local people.

Figure 8.2 National forest conservation area

Phase 3

As a logical third step, there needs to be clearer acknowledgment in national forest policy statements of the importance of on-farm tree resources and remnant woodlands. In most developing countries these provide a very high proportion of essential fuelwood, fodder, fruit, and other products. They play a key role in agricultural sustainability, in food security, in generation of rural incomes, and in ensuring a supply of raw material for small-scale rural enterprises.

Historically, quantification of these on-farm tree resources and acknowledgment of their importance has largely gone unrecognized in the forest policies of many countries. Explicit recognition in national forest strategies of the importance of on-farm tree resources will ensure allocation of adequate resources for agroforestry, education, training, research, and extension programs.

Setting Standards for Monitoring the Permanent Forest Estate

Since the Rio Earth Summit there has been increased focus on the development of criteria for and indicators of sustainable forest management, as well as a growing acceptance of the principles of target setting, monitoring, and accountability of forest agencies and national governments for progress toward sustainable conservation and management of forests. These standards are necessary for monitoring both forest management and trade in forest products. The former need to reflect social, technical, economic, and environmental objectives of sustainable forest management. Standards for regional and international trade in forest products are likely to be increasingly subject to the certification and "ecolabeling" procedures that are currently being adopted in many developed and developing countries.

The usefulness and effectiveness of such standards will depend to a high degree on the extent to which local communities and—in the case of industrial forests—the private sector have been fully involved in their formulation. The development of timber certification and ecolabeling practices will, in some situations, need to involve overseas consumers in an assessment of their willingness to pay a premium for products derived from forests that have been certified as under sustainable management. Recognition of these potential linkages between national producers and overseas consumers needs to be reflected in national forest policy statements.

Defining Different Stakeholder Roles

The definition and sharing of various stakeholder roles in sustainable conservation and development of forest resources will vary considerably between countries. Recent history has, however, seen an emerging consensus on the desirability of a significant reallocation of the responsibilities of governments, local communities, and the private sector. National forest policies could usefully reflect this trend.

Generally speaking, national governments may need to retain primary responsibility for policy development, law enforcement, development of standards, legislative regulations, and environmental protection. Some government forestry institutions are moving toward playing a supportive role in research and extension (particularly in the agroforestry arena), rather than continuing to devote their main effort to management of the government-owned forest estate.

Farmers, indigenous communities, and women's groups are increasingly being acknowledged as the primary "managers" of forest resources. This reflects recognition that a high proportion of local people's forest product needs—both for subsistence and income generation—are being derived from farm trees and remnant woodlands rather than from large blocks of continuous forest.

The private sector (and particularly forest industries) are being called on to assume a more prominent role in the areas of investment in and management of industrial forests (particularly of commercial-scale industrial plantations).

The role of NGOs (which was rarely recognized in earlier national forest policy statements) is now well established. They have key functions to perform in raising awareness of social and environmental problems, in stimulating policy dialogue, and in some situations acting as brokers of more effective partnerships between government agencies, local communities, and concerned conservation groups.

More explicit recognition of the respective roles of various stakeholders in national forest policy statements would go a long way toward eliminating the conflicts and tensions that have characterized recent forest policy debate.

Resource Assessments and the Monitoring of Forest Change

As thinking about the role of forests and of various stakeholders has evolved in recent years, so also have the functions and responsibilities of government agencies involved in forest resource assessments. Historically the main focus was on timber inventories and sustainable timber management. Today there is also a strong emphasis on the need for improved inventorying of the vast array of nontraditional forest products that play a role in meeting local community subsistence needs, many of which also make a major contribution to international trade.

More recently, rising international concerns about tropical deforestation and the declining quality and health of many temperate and boreal forests have triggered regional and global forest resource assessments aimed at monitoring global forest change, assessments focused in particular on the status of the forests' biodiversity and their effects on climate.

National forest policies of the future will have to provide for such forest resource assessment and resource-monitoring activities. One element deserving of special attention is the possibility for achieving improved national, regional, and global harmonization in collation, publication, and dissemination of data and resource information. The current proliferation of resource assessment methodologies and activities, combined with the incompatibility of many geographic information systems, is resulting in major inconsistencies in reporting and difficulties in securing reliable data as the basis for policy recommendations.

An emerging area of concern is the need for effective assessment mechanisms for monitoring the commitments now being made by many national governments to adoption of the requirements of global conventions such as those covering biodiversity and climate change, as well as the various emerging regional forest agreements and conventions, such as those developed through the Helsinki and Montreal Processes, the Central American Forest Convention, and emerging informal agreements in the Amazon, the South Pacific, and other regions.

Implications for National Capacity-Building

The above analysis suggests that the elements of an effective integrated national forest strategy will likely include

1. policies for sustainable development and conservation of forests that adequately reflect the interests of different stakeholders;
2. harmonization of policies in closely related sectors such as agriculture, transportation, energy, and industry;
3. appropriate emphasis on key areas of policy reform and institutional change that are needed to provide an enabling framework for sustainable forest conservation and development;
4. clear identification of the major issues and prioritization of activities with a balanced focus on the poverty-alleviation, trade, and environmental protection aspects of sustainable forest management;
5. setting of standards for the use and management of forests;
6. reappraisal of earlier approaches to forest land-use planning and a willingness to reconsider traditional forest boundaries;
7. recognition of the linkages between national and regional and international forest policies and in particular of the commitments being made by national governments under various regional and international conventions;
8. appropriate mechanisms for assessment and monitoring of forest resource change and of national progress in responding to the requirements of regional and international conventions.

Effective translation of these various elements of an integrated national forest strategy into sustainable, action-oriented programs will require adequate allocation of human and financial resources to national capacity-building.

Implications for Forestry Education and Training

The new approaches to the formulation of national forest strategies outlined above have significant implications for both training curricula and modes of forestry education (see Chapter 5). Increasing emphasis will be needed, for example, on the sociological and cultural dimensions of forest policy formulation. Government forest agencies will need to be better trained in the areas of agroforestry-related research needs and rural extension techniques. By definition, this is likely to have major implications for gender-based work patterns (a high proportion of developing country households and farms are managed by women who play the main role in household decisions on local tree harvesting and management; figure 8.3).

Figure 8.3 Extension officer meeting with village women

Forest education and training will need to factor in the important cross-sectoral linkages that are key to sustainable land-use management and preservation of the environmental roles of forests. More emphasis than in the past will need to be devoted to improved understanding of broadly based institutional mechanisms that provide a basis for interactive national forest policy planning.

Participatory National Policy Formulation

There is now quite a long history of community participation in agriculture as well as forestry development, and a wide range of development agencies (both national and international) have attempted to involve local people in planning of conservation and development programs. A failing of many such attempts in the past was an overly simplistic assumption that local participation would be assured merely by allowing local communities to comment on government-formulated forest plans and obtaining their apparent consent.

There is much that forestry professionals can learn from agricultural experiences in this area. In the agriculture sector many varying approaches and participatory methods for learning and action have been tested and systematically evaluated. Analyses of these experiences suggest at least seven different types of participatory approaches (Pretty 1995):

1. *Passive participation.* People participate by being told what is going to happen or has already happened. It is a unilateral announcement by an administration or by project management without listening to people's responses. The information being shared belongs only to external professionals.
2. *Participation in information giving.* People participate

by answering questions posed by extractive researchers using surveys or similar approaches. People
do not have the opportunity to influence proceedings,
since the findings are neither shared nor checked for
accuracy.

3. *Participation by consultation.* People participate by
being consulted, and external agents listen to people's
views. These external agents define both problems
and solutions and may modify these in the light of people's responses. Such a consultative process does not
concede any share in decision making, and professionals are under no obligation to take people's views
on board.

4. *Participation for material incentives.* People participate by providing resources, for example, labor, in return for food, cash, or other material incentives. Much
on-farm research falls in this category as farmers provide the fields but are not involved in experimentation or in the process of learning. It is very common to
hear this called participation, yet people have no
stake in prolonging activities when the incentives
end.

5. *Functional participation.* People participate by forming groups in order to meet predetermined objectives
related to the project, which can involve the development or promotion of externally initiated social organization. Such involvement does not tend to be at early
stages of project cycles or planning, but rather after
major decisions have been made. These institutions
tend to be dependent on external initiators and facili-
tators, but may become self-sustaining.

6. *Interactive participation.* People participate in joint
analysis, which leads to action plans and the formation of new local institutions or the strengthening of
existing ones. This approach tends to involve interdisciplinary methodologies that seek multiple perspectives and make use of systematic and structured
learning processes. These groups take control over lo-

cal decisions, and so people have a stake in maintaining the structures or practices.

7. *Self-mobilization.* People participate by taking initiatives independent of external institutions in order to change systems. They develop contacts with external institutions for resources and technical advice they need but retain control over how resources are used. Such self-initiated mobilization and collective action may or may not challenge existing inequitable distributions of wealth and power.

As forest policy formulation moves toward more interactive approaches with local communities, of paramount importance will be the willingness of forest agencies to develop sensitivity and the capability for learning from local forest-dwelling communities about their perceptions of the uses and value of forest resources. Indigenously derived approaches to forest harvesting, management, and conservation will, in many situations, be a more logical starting point for policy formulation than imported technologies based on increasing the productivity of selected species.

The professional forester's role in all of this is best interpreted as that of a facilitator of local people's analyses of their needs for institution-building. The goal is to return responsibility for management of state-owned forest resources to the local residents. In short, the forester's job is to strengthen forest dwellers' capacity to take action on their own. Successful incorporation of such approaches into national capacity-building programs will largely determine the longer-term sustainability of forest resource management.

Cross-Sectoral Linkages

Despite more than a decade of rhetoric on the need for forestry agencies to become more outward-looking and to develop closer integration with other sectoral agencies, tangible progress toward developing effective institutional mechanisms for achieving this goal has remained elusive.

Earlier approaches by the Tropical Forestry Action Programme (TFAP) were characterized by an excessive focus on forest-sector institutions and forestry programs. More recéntly the TFAP has begun to broaden its horizons. There is growing recognition of the desirability of integrating forest concerns into the analyses and recommendations of national environmental action plans, national conservation strategies, and national sustainable development strategies.

Similarly, forestry agencies will need to become more proactive in trying to ensure that the policies of other sectors such as agriculture, transportation, energy, industry, and mining factor forest conservation concerns into their own sectoral policies.

Linkages Between National, Regional, and Global Institutions

Globally agreed "rules of the game" for forest management such as the Agenda 21 Forest Principles and forest-related obligations of the various emerging regional conventions state the need for a special focus in national capacity-building on appropriate methodologies and institutional mechanisms for monitoring performance.

Many such exercises are already under way. Advantage is being taken of remote sensing, geographic information systems, and other technologies for monitoring of forest change and for collation and dissemination of data relating to the status of national forests and progress toward accepted targets.

*Mechanisms for Encouraging Policy Reform and
Conflict Resolution*

Uncertainty about the likely impact of policy reforms has
been one constraint on implementation of such reforms. In
national forest policy statements there needs to be more explicit recognition of the potential of specialized policy research institutions and universities involved in forest policy
analysis to contribute to better understanding of the possible
impact of policy reforms.

Other options for encouraging more positive action by national governments to address reforms and to work toward
conflict resolution include enlisting the support of NGOs in
raising awareness of the consequence of failure to implement
such reforms, as well as support for regional policy seminars
to encourage transparent dialogue between various stakeholders and to disseminate experiences of policy reform and
collaboration with international agencies so as to address
sensitive issues in which conflict resolution could benefit by
study of overseas experiences.

The development banks have a special role to play in the
policy reform arena. Through carefully designed conditional
lending and structural adjustment operations they can help
weak forest administrations to secure support for policy reforms that will have a positive impact on forests. Conditional
lending approaches are most likely to be successful in situations where weak forest administrations lack the political
clout to influence national political policies and development strategies.

Funding Implications and Coordination

Implementation of a more broadly based approach to formulation of integrated national forest strategies as outlined

in this chapter clearly has significant funding implications. Special effort will be needed to mobilize incremental resources. The UNDP Capacity 21 program could play a leading role in this area.

Mobilization of increased financial support for capacity-building is likely to be enhanced by the current trend toward adoption of natural resource accounting systems, which factor the costs of forest depletion into national accounts. Such approaches will help to heighten political awareness of the important contribution that forests make to both economic development and environmental protection.

Achieving improved aid coordination will remain a challenge. The past activities of the Forestry Advisers Group have been helpful in this regard. Prospects for improved aid coordination in the future could be further enhanced by a strong focus on removing existing constraints to delivery of aid-supported programs through national channels. This implies the need for wider stakeholder participation in aid-policy dialogue and greater accountability, transparency, and efficiency in the activities of national institutions.

Summary

This chapter stresses the importance of factoring into national policies the global and regional roles and benefits of forests. It places special emphasis on the need for alternative approaches to national forest strategy formulation. From the outset, it should be recognized that different stakeholders have different perceptions of what constitutes sustainable forest management. Key elements of an integrated national forest strategy are presented. The importance of prioritization of key goals is discussed, as is reform of both regulatory and financial instruments. The three phases of an approach to developing a forest strategy that is quite different from traditional ones are set out. Techniques for monitoring and for redefining the role of local stakeholders are presented, as are the implications of this for national capacity-building. The nature of people's participation is examined, leading to insights on the key role of a cross-sectoral framework. Finally, the implications of this for funding support, cooperation, and partnership are explored.

CHAPTER NINE

The Problem and Potential Solutions
A Summary

JOHN C. GORDON, JOYCE K. BERRY, AND
RALPH SCHMIDT

The Problem

The greatest lesson from the past regarding forest strategies is that deforestation remains a severe and accelerating problem despite substantial efforts to reduce it. Past programs such as TFAP and many country initiatives and plans have brought a much better understanding of the ingredients of potentially successful forest strategies, however, and they have inculcated into the development assistance community a sensitivity to the need for initiative and leadership to lie with countries rather than with international or bilateral agencies. At the same time, most developing countries stress cooperation with the international and bilateral agencies as a critical element in successfully implementing their own strategies. Detailed knowledge of institutions and programs, existing and proposed, is necessary if countries are to obtain maximum assistance in creating integrated forest development strategies. This information is scattered and difficult to interpret. International conventions, agendas, and movements are coalescing around deforestation, biological diver-

sity, and global change. All will directly or indirectly influence forest management and forest dwellers everywhere. At
the same time, local institutions and the developers of
ecosystem management are focusing on local participation
and control, human values, the specific properties of specific
ecosystems, and techniques and organizations closely
adapted to local values, capabilities, and conditions. International and local directions and methods must be recon-
ciled and harmonized if progress toward locally sustainable
forestry is to be realized.

The Solution

Organizations can be most effective in fostering development and combating deforestation by assisting in the creation of integrated forest strategies focusing on poor farmers
and forest dwellers. These should be created and led "in-
country" and funded through carefully crafted partnerships
with a spectrum of development assistance groups appropriate to the country and to the specific problems at hand.

Poverty is rightly viewed as one cause of deforestation, although as Pearce points out in Chapter 7 the quantitative evidence for this is equivocal. With no better alternatives, forest destruction and degradation become a short-term
solution to many of the burdens imposed by poverty. Again,
as Pearce points out, the rise of incomes from the lowest levels may initially increase forest-clearing, and only policies
and markets that reward working of existing cropland will
abate this clearing.

It is equally true, however, that for poor people who live
in and near forests, and those who live in places where
forests have been destroyed or degraded, forests and trees
can be effective in reducing poverty and ameliorating its ef-

fects on people and the landscape, as Benbrook and Peluso point out in Chapters 3 and 4.

Forests, agroforestry, and tree crops can provide increased and sustained employment in the production and processing of timber. Because of their perennial habit, they can be sources of stable production and capital formation through drought and other periods of climatic and market adversity. Unlike annual crops, wood harvest can often be delayed to coincide with need or better market conditions. Trees and forests, by providing shade, fodder, soil stabilization, and good watershed conditions, can improve quality of life for those struggling to escape poverty and provide for many of the spiritual and aesthetic needs of local people.

It is thus as important to view trees and forests as effective potential tools for poverty alleviation as it is to understand the connection between poverty and deforestation. This view of forests as development assets particularly relevant to the poor is sometimes viewed as antienvironmental. But most now agree that unless poverty is alleviated in and around forests, they will continue to be lost. It should also be clear that the sustainable use and reestablishment of forests is the closest tool (in both time and space) to use in fighting poverty in and near forests.

Many opportunities exist to use specific forestry and agroforestry practices to enhance the economic and spiritual well-being of local people. The resources to do these on a large scale, however, and the coordination required to avoid major errors and crossed purposes, will only arise from careful and integrated planning. Thus the steps to using forests as a poverty-fighting tool are, as indicated in Chapter 2,

1. Recognize at the national and local levels that forests are a primary poverty-fighting asset and begin to view and describe them in that way;

2. Create an integrated national strategy for forests with this view at its heart;
3. Use the strategy to create partnerships that produce the resources for large-scale action and that reduce the current fragmentation of effort;
4. Build institutional capacity by creating knowledge and expanding research, education, and outreach; and
5. Implement the strategy with broad participation and careful monitoring and feedback in an "adaptive management" approach.

Poor rural families in and near forests and resource-poor people who can migrate to them are thus the key to successful sustainable development of forests, as well as to the eventual reduction of deforestation rates and the regeneration of woodlands. Only if their condition, goals, and motivations are understood can successful, integrated strategies be created. These will differ from country to country, but there are common elements and themes. For example, the role of women is almost always crucial in forest development strategies (Patel-Weynand 1996), as are patterns of migration, the forms of agriculture practiced, and the availability of credit. In areas of moist tropical forest where human development levels are relatively high (peninsular Malaysia, Queensland, Puerto Rico) marginal shifting of cultivation is abandoned and forest management is possible. The minimum needs of rural poor families include basic education, health services, and economic opportunity if they are to avoid unsustainable agriculture and deforestation. Meeting these needs is also a key to population stabilization, which usually plays an equally critical role in sustainability.

Well-managed forests can better sustain the livelihood of resource-poor people in and around forests, but management must be carefully tailored to specific human and forest conditions (for example, through "ecosystem management"), and the people themselves must participate fully in decisions made in pursuit of better management. Access, ownership, and property rights issues are often the determinants of whether forests sustain or degrade the livelihood of smallholders. Forest management is always embedded in a larger context of culture, politics, and trade, as are the smallholders themselves. In this larger context, forests provide a useful metaphor for sustainable development on a larger scale because of their complexity, longevity, and multiple outputs and benefits, the latter often crucial to the sustenance of the poorest of the poor. Forests also link to all other sectors and economic factors, especially agriculture (including livestock and range), energy, and employment generation and allocation.

Education is a key to all levels of activity in sustainable forestry. Informal or noninstitutional education is particularly important for smallholders and forest-dwellers. The development of usable technology and an explanatory and predictive science base are particularly necessary for ecosystem management leading to sustainable forest use and development. An interdisciplinary approach that includes biological, environmental, economic, cultural, and social research and that produces integrated answers to field-generated questions must be developed specifically for the conditions and goals of the country producing the integrated forest strategy. Rapid technology transfer and clear communication between researchers and forest owners and operators is as necessary as the research base.

Knowledge of the true costs and benefits of forest-related human activities is necessary to judge the effectiveness of ecosystem management. Adaptive (learning) management can only be practiced if the consequences of management plans and decisions are accurately estimated "on the ground." We now know that in order to do this, it is necessary to account for the depletion of natural capital and for "externalities" previously not included in economic assessments of forest harvesting and other activities. We need to apply this "new" economics at all levels, from the local to the international, in order to understand and implement fully integrated forest strategies.

Integrated forest strategies (IFSs) are tailored to specific country objectives and conditions using the principles of ecosystem management and adaptive management. Thus, there is no single recipe or format for creating them. Most IFSs will have common elements, however, and all use the same set of general principles. For example, a nation's developing a countrywide vision of the future of its forests is a necessary first step toward the simultaneous reduction of deforestation and achievement of sustainable forest development. The methods used to develop the vision will vary with the culture and conditions of individual countries, but the vision will not be solely developed by specialists and will reflect widely supportable ideas about the forests' role in the larger context of development and environmental protection. Creating an integrated forest strategy requires that the interactions of the forest with other sectors be understood and that all important beneficial aspects of forests be recognized. Especially, it requires a holistic view of the forest ecosystem and its contents, inputs, and outputs, particularly including humans and their needs and impacts.

The formulation of a countrywide vision for forests in the context of the country's overall development strategy—and the careful definition of problems and opportunities specific to the country—are at the core of an integrated forest strategy. All other components of a strategy flow from these. A summary list of the contents of a forest strategy includes

- Forest characteristics and social, economic, and political characteristics
- A countrywide vision
- Evidence of political commitment
- Development and environmental goals
- Description of the fabric of achievement:
  - Problems and opportunities
  - Alternative solutions and initiatives
  - Assessment of consequences of alternative solutions
  - Internal agreement protocols (among sectors, agencies, levels of government, geographic areas, communities)
  - External agreement protocols (among countries, international development assistance agencies, trading partners)
- Adaptive management structure:
  - Objectives, responsibilities, organizational structures, targets
  - Information needed to describe progress
  - Feedback channels to adjust goals and methods (monitoring)
  - Public participation and conflict resolution
- A communication plan and implementation structure.

Whether this process will go forward will depend on the formation of partnerships, particularly between the developed and the developing countries. There is a particular problem related to the creation of these partnerships with which we close. Tropical forests present two contrasting faces to the developed world (and to important parts of the developing world as well). The first is usually quite clearly seen: a picture of forests pristine but threatened, magnificent but precarious, ecologically and emotionally necessary, and in need of "saving." The second is usually seen dimly, if at all; a cynic would say that even the will to see it is suppressed. This picture frames people in and around the forests to be "saved"; they depend on the forest in every way for a usually meager livelihood. They must use the forest, and in its richness lies their only current hope of discontinuing their membership in that celebrated but poorly understood club, "the poorest of the poor." The people in the second picture are found wherever tropical forests are found in Asia, Africa, and Latin America. They are counted not in millions or tens of millions but in hundreds of millions.

Unless these separate views are forged into a common vision for action, there is little hope for either the forests or the people, and perhaps little hope for the moral, ecological, or political salvation of the developed world. The first step in forming such a vision is to gain a clearer view of the problem in order to place it before those who might have the money, will, and motivation to solve it. Clearly, this will include primarily the people in and near forests and the national governments and public and private institutions that contain them. But it must also include the hearts, minds, and pocketbooks of the developed world and particularly its development and financial institutions. The second step is to generate methods that would underlie an agenda for action at the national and local levels. If a single lesson stands out from past efforts, it is that unless local people are committed and providing leadership, forest initiatives either fall flat or slowly erode away.

This book is intended to help those intent on making those first two critical steps. Of course, no single book will suffice. We hope to raise the level of the debate, accelerate learning (a clear lesson is that we don't know much yet in comparison to the size of the problem), and sound a reasoned call to action. Action will be difficult and complex. It will be far removed, for the most part, from most current glamorous "conservation and environment" activities. Thus, for example, it will have little to do with sending helicopter gunships after poachers or counting the species in the Amazon. Much of it will take place in the communities of the poorest of the poor. Great effort must go into putting the forest at their service while maintaining its vitality. But much of the most critical action will take place around expensive tables (usually made from trees from the forests in question), in cities far removed from the forest, where major investment decisions are made. The role of science will be more to solve problems related to both using and keeping the forest than to further describe its wonders. In short, great change will be required of all who now think they are engaged in either using or saving tropical forests.

Key Questions Answered

Vast areas of unknown intellectual territory lie between the present and the eventual creation of fully effective country forest strategies. Nevertheless, we have provisional answers to each of the key questions raised in the second chapter of this book. Time and their practical employment will bring continued improvement.

1. Who lives in and around forests, and how do they interact with the forests, with each other, and with more remote actors, such as governments and businesses?
   - A forest is a social and political place.
   - People's relations with other people are as important to understanding their use of the forest as are people's direct forest management activities.
   - Poorer people with little direct access to wealth-producing opportunities can expand their social networks and thereby increase the number of channels

through which they can acquire forest resources. The more networks and people involved, the greater the pressure on the forest.
- State forest policy and international influences need to be understood as contemporary means of allocating access to resources—or property rights—which may change the distribution of benefits from forest use.
- A political ecology approach to analyzing the ways forests sustain people's livelihoods and the ways these livelihoods are threatened can help identify critical processes, structures, and relations affecting forest use.
- To do political ecology:
    1. Start by identifying resource users and particular resource management practices or strategies, institutions for resource access and control, and sources of conflict and cooperation.
    2. Examine relevant state and market systems or aspects of forest management superimposed on local or customary systems for the production of timber and other local products.
    3. Examine international influences and the reverberations of local activities "outward" to the national and international levels.
    4. Assess unanticipated outcomes from the interactions between multiple resource strategies.
    5. Think of local, state, and international systems of management as a set of cumulative impacts on the ways individual people actually use forests.
2. What institutional settings and approaches can be identified from past efforts to aid in the creation of new country strategies?
    - National forest planning, programming, and project implementation have occurred, although the scale is still minor compared to the economic and human population forces involved.
    - Lack of acceptable results may be attributed to this

scale, but so may the real conflicts of interest within affected countries and among forested countries and rich nations that contribute to international cooperation.

- New forest programs will face the daunting challenge of requiring strong high-level political support *and* broadening poor people's participation.
- In many countries, international collaboration is essential to confront the problems, but current patterns of global cooperation are totally inadequate for successful outcomes.
- International action on forests takes place within a thirty-year-old system of international agreements on global environment and development issues and institutions to support agreed actions.
- Forest policy can be no more coherent or effective than land-use policy overall.
- Many of the most promising developments occur when forest management is placed in the hands of the local community.

3. What human resources and what capacity-building activities are necessary for practical pursuit of new goals using new strategies?
   - Knowledge is not only the source of the highest-quality power but also the most important ingredient of force and wealth.
   - Perhaps the most vital difference between developed and developing, rich and poor, is the gap between these groups in the capacity to generate, acquire, disseminate, and use scientific and technological knowledge.
   - Empowerment through participation involves the recognition that local people are uniquely suited to decide what is best for their development. Thus, participation, with appropriate knowledge, should be the basis for development.
   - Training and education should include
     - Strengthened attitudes and values

- Formal and informal education
- Socioeconomic and cultural aspects
- Balance between practical skills and theoretical knowledge
- Understanding of the major problems and opportunities existing in forestry development
- Motivation to work in rural forested areas
- Research needs
  - Government policy and legislation to encourage science and technology generation
  - Tapping of knowledge generated at regional and international centers
  - Linkages between new knowledge and practical local knowledge
  - A critical mass of scientists
  - Strong leadership
  - An outreach component to communicate results
  - Understanding that research is an investment, not an expense
  - A focus on understanding ecosystems, people and cultures, and economics
  - Management practices and options
  - Fiscal and financial resources
- Outreach should
  - Disseminate and transfer research results, science and technology, and experience
  - Strengthen institutions
  - Provide feedback to researchers
  - Include NGOs, private industry, and entrepreneurs

Interactions between education and training, research and validation, and outreach and technical cooperation must exist in order for sustainable forestry development to occur.

4. What are the economic drivers of deforestation, and how can new methods, particularly of valuation, be used as a basis for strategy development?
   - Securing the sustainable management of tropical

forests will not be achieved by simply declaring an area to be protected. A closer look at the underlying structure of incentives for forest destruction is needed. The following areas should be focused on for maximum exploitation of sustainability:

- Prevailing distortions: subsidies, concessions, tax regimes, infrastructure construction
- Land tenure and resource rights to lower discount rates
- Differential credit availability and price, also affecting the discount rate
- Productivity-increasing ventures on already converted land
- "Global bargains," either through the GEF or through the attraction of private-sector deals involving biodiversity prospecting and carbon offsets.

The end result is a package of measures generated by what some call a new economics but what in reality is an application of well-tried economic theories.

5. What are the components of an effective country strategy, and how can they be integrated and implemented?

National policies must reflect the global and regional roles and benefits of forests. The greatest need is for alternative, more broadly based and local-people-oriented approaches to national forest-strategy formulation. From the outset, it should be recognized that different stakeholders have different perceptions of what constitutes sustainable forest management, and all must be served in policy formulation. Techniques for monitoring and for redefining the role of local stakeholders are key, as are the implications of this for the creation of greater human capacity to manage forests in the broadest sense.

6. How do forest strategies connect with economic and social activities traditionally considered to lie in other sectors (agriculture, energy, manufacturing, international trade)?

The essence of practical country strategy making is to address specific sectors such as forestry in relatively full knowledge of all forest components and values, as well as those of other sectors and activities (such as agriculture and energy development), but on a scale that recognizes local and fine-grained influences and that stops well short of completely homogenized and universal grand plans. The major requirements for doing this are

- A greatly improved knowledge of forests and human values based on better inventories of both;
- More effective cross-sectoral information exchange; and
- Effective coordination between local and national goals and methods.

Bibliography

Agarwal, B. 1986. *Cold Hearths and Barren Slopes: The Woodfuel Crisis in the Third World.* Riverdale, Md.: Riverdale.

Allen, J., and D. Barnes. 1985. The causes of deforestation in developing countries. *Annals of the Association of American Geographers* 75 (2).

Berry, S. 1989. Social institutions and access to resources. *Africa* 59 (1): 41–55.

Blaikie, P. 1985. *The Political Economy of Soil Erosion in Developing Countries.* London: Longman.

Blaikie, P., and H. Brookfield. 1987. *Land Degradation and Society.* London: Methuen.

Brown, K., and D. W. Pearce, eds. 1994. *The Causes of Tropical Deforestation: The Economic and Statistical Analysis of Factors Giving Rise to the Loss of the Tropical Forests.* London: University College Press; and Vancouver: University of British Columbia Press.

Burgess, J. 1992a. *Economic Analysis of the Causes of Tropical Deforestation.* London Environmental Economics Centre, Discussion Paper 92–03, November 1992.

———. 1992b. Economic analyses of frontier agricultural expansion and tropical deforestation. M.Sc. diss., University College, London.

Capistrano, A., and C. Kiker. 1990. *Global Economic Influences on Tropical Broadleaved Forest Depletion.* Washington, D.C.: World Bank.

CATIE. 1994. *Agenda for a Critical Decade: Strategic Plan 1993–2002.* Turrialba, Costa Rica: CATIE.

Chambers, R., and M. Leach. 1989. Trees as savings and security for the rural poor. *World Development* 17 (3): 329–42.

Constantino, L., and D. Ingram. 1990. *Supply-Demand Projections for the Indonesian Forestry Sector.* Jakarta: FAO.

Cuesta, M., G. Carlson, and E. Lutz. 1994. *An Empirical Assessment of Farmers' Discount Rates in Costa Rica and Its Implication for Soil Conservation.* Washington, D.C.: World Bank.

Deacon, R. 1994. *Deforestation and the Rule of Law in a Cross Section of Countries.* Resources for the Future (Washington, D.C.), Discussion Paper 94–23.

English, J., M. Tiffen, and M. Mortimore. 1994. *Land Resource Management in Machakos District, Kenya, 1930–1990.* Environment Paper no. 5. Washington, D.C.: World Bank.

Eponou, T. 1993. *Partners in Agricultural Technology: Linking Research and Technology Transfer to Serve Farmers.* Research Report 1. The Hague: International Service for National Agricultural Research.

Fankhauser, S. 1995. *Valuing Climate Change.* London: Earthscan.

FAO. 1995a. *World Agriculture: Toward 2010.* Rome: FAO.

FAO. 1995b. *Report of the Committee on Forestry.* 12th sess. Rome: FAO.

Fernow, B. 1911. *The History of Forestry.* 3d ed. Toronto: Toronto University Press.

Fortmann, L. P. 1990. Locality and custom: Non-aboriginal claims to customary usufructuary rights as a source of rural protest. *Journal of Rural Studies* 6 (2): 195–208.

———. 1985. The tree tenure factor in agroforestry with particular reference to Africa. *Agroforestry Systems* 2: 229–51.

Godoy, R., and R. Lubowski. 1992. Guidelines for the economic valuation of nontimber tropical-forest products. *Current Anthropology* 33 (4): 423–33.

Gordon, J. 1994. Vision to policy: A rule for foresters. *Journal of Forestry* 92 (7): 16–19.

Grove, R., and D. Anderson, eds. 1989. *Conservation in Africa: People, Policies, and Practice.* Cambridge: Cambridge University Press, 1987.

Guevara, R., and A. J. Valle. 1988. Educating forestry professionals for rural development needs. In *Proceedings of the Regional Conference on Education and Training of Agriculture Professionals for Rural Development.* Edited by E. Valdivia. Mexico City: Collegio de Postgraduados and FAO Mexico.

Guha, R. 1990. *The Unquiet Woods: Ecological Change and Peasant Resistance in the Himalayas.* Berkeley: University of California Press.

Gurderion, L. H., C. S. Hollings, and S. Light, eds. 1995. *Barriers and Bridges to the Renewal of Ecosystems and Institutions.* New York: Columbia University Press.

Hafner, J. A. 1990. Forces and policy issues affecting forest use in northeast Thailand 1900–1985. In *Keepers of the Forest: Land Management Alternatives in Southeast Asia,* edited by M. Poffenberger. West Hartford, Conn.: Kumarian.

Hall, B. K. 1985. *Maritime Trade and State Development in Southeast Asia.* Honolulu: University of Hawaii Press.

Hardin, G. 1968. The tragedy of the commons. *Science* 162: 1243–48.

Hecht, S., and A. Cockburn. 1989. *The Fate of the Forest: Developers, Destroyers, and Defenders of the Amazon.* London: Verso.

Hirsch, P. 1990. *Development Dilemmas in Rural Thailand.* New York: Oxford University Press.

IDRC. 1991. *Empowerment Through Knowledge: The Strategy of the International Development Research Center.* Ottawa: IDRC.

IFAD. 1992. *The State of World Rural Poverty.* Rome: IFAD.

IPF. 1996. *Report of the Intergovernmental Panel on Forests.* New York: UNDP.

Kahn, J., and J. McDonald. 1994. International debt and deforestation, in Brown and Pearce (1994).

Katila, M. 1992. Modelling deforestation in Thailand: The causes of deforestation and deforestation projections for 1990–2010. Finnish Forestry Institute, Helsinki. Mimeographed.

Kramer, R., E. Mercer, and N. Sharma. 1994. Valuing tropical rain forest protection using the contingent valuation method. School of the Environment, Duke University, Durham, N.C. Mimeographed.

Kummer, D., and C. H. Sham. 1994. The causes of tropical deforestation: A quantitative analysis and case study from the Philippines, in Brown and Pearce (1994).

Leach, T. A. [1919] 1988. Date-trees in Halfa Province. In *Whose Trees? Proprietary Dimensions of Forestry*, edited by L. P. Fortmann and J. W. Bruce. Boulder: Westview.

"Limits to Growth." 1972. Abstract established by Eduard Pestel from a report to the Club of Rome, by Donella H. Meadows et al.

Macpherson, C. B. 1983. *Property: Mainstream and Critical Positions*. Toronto: University of Toronto Press.

Mahar, D. J. 1989. *Government Policies and Deforestation in Brazil's Amazon Region*. Washington, D.C.: World Wildlife Fund and the Conservation Foundation.

Meiggs, R. 1982. *Trees and Timber in the Ancient Mediterranean World*. Oxford: Clarendon.

Moreddu, C., K. Parris, and B. Huff. 1990. Agricultural policies in developing countries and agricultural trade. In *Agricultural Trade Liberalization: Implications for Developing Countries,* edited by I. Goldin and O. Knudsen. Paris: OECD.

National Research Council. 1990. *Forestry Research: A Mandate for Change*. Washington, D.C.: National Academy of Sciences Press.

Neumann, R. 1992. The political ecology of wildlife conservation in the Mount Meru area of Tanzania. *Land Degradation and Rehabilitation* 3: 85–98.

OECD. 1993. *Agricultural Policies, Markets, and Trade: Monitoring and Outlook*. Paris: OECD.

Palo, M., G. Mery, and J. Salmi. 1987. Deforestation in the tropics: Pilot scenarios based on quantitative analyses. *Metsatutkimuslaitoksen Tiedonantaja* 272 (Helsinki).

Panayotou, T., and S. Sungsuwan. 1994. An econometric study of the causes of tropical deforestation: The case of northeast Thailand. In Brown and Pearce (1994).

Patel-Weynand, T. 1996. *Forests and Poverty: A View Towards Sustainable Development.* New York: UNDP.

Pearce, D. W. 1995. Global environmental value and the tropical forests: Demonstration and capture. In *Forestry and the Environment: Economic Perspectives II,* edited by V. Adamaovicz. London: CAB. Forthcoming.

Pearce, D. W., and D. Moran. 1994. *The Economic Value of Biodiversity.* London: Earthscan; and Washington, D.C: Island, in association with IUCN.

Pearce, D. W., and J. Warford. 1993. *World Without End: Economics, Environment, and Sustainable Development.* Oxford and New York: Oxford University Press.

Peluso, N. 1995. Whose woods are these? Counter-mapping forest territories in Kalimantan. *Antipode* 27 (4): 383–406.

———. 1993. Coercing conservation: The politics of state resource control. *Global Environmental Change* 4 (2): 199–217.

———. 1992a. The ironwood problem: (Mis)-management and development of an extractive rainforest product. *Conservation Biology* 6 (2): 210–19.

———. 1992b. *Rich Forests, Poor People: Resource Control and Resistance in Java.* Berkeley: University of California Press.

———. 1992c. The political ecology of extraction and extractive reserves in East Kalimantan, Indonesia. *Development and Change* 23 (4): 49–74.

Peluso, N., C. R. Humphrey, and L. P. Fortmann. 1994. The rock, the beach, and the tidal pool: People and poverty in natural resource–dependent areas. *Society and Natural Resources* 7: 23–28.

Peluso, N., M. Turner, and L. P. Fortmann. 1994. *Introducing*

Community Forestry: Annotated Listing of Topics and Readings. Rome: FAO.

Perrings, C. 1992. An economic analysis of tropical deforestation. Environment and Economic Management Department, University of York, UK.

Peters, C. 1994. *A Practitioner's Guide to Tropical Forest Ecology.* Washington, D.C.: World Wildlife Fund Biodiversity Support Network.

———. 1991. Environmental assessment of extractive reserves: Key issues in the ecology and management of non-timber forest resources. Draft report to Environment Department, World Bank, Washington, D.C.

Peters, C., A. Gentry, and R. Mendelsohn. 1989. Valuation of an Amazonian rainforest. *Nature* 339: 655–56.

Peterson, R. B. 1991. To search for life: A study of spontaneous immigration, settlement, and land use on Zaire's Ituri Forest frontier. Master's thesis, University of Wisconsin, Madison.

Phillips, D. 1994. The potential for harvesting fruits in tropical rainforests: New data from Amazonian Peru. *Biodiversity and Conservation* 2: 18–38.

Pretty, J. N. 1995. *Regenerative Agriculture: Policy and Practice for Sustainability and Self-Reliance.* Washington, D.C.: Joseph Henry.

Reid, A. 1988. *Southeast Asia in the Age of Commerce.* Vol. 1, *The Lands Below the Winds.* New Haven, Conn.: Yale University Press.

Reis, J., and R. Guzman. 1992. An econometric model of Amazon deforestation. Paper presented at conference, Statistics in Public Resources and Utilities and in the Care of the Environment, Lisbon, April 1992.

Repetto, R., and M. Gillis. 1988. *Public Policies and the Misuse of Forest Resources.* Cambridge: Cambridge University Press.

Rudel, T. 1989. Population, development, and tropical deforestation: A cross-national study. *Rural Sociology* 54 (3). Reprinted in Brown and Pearce (1994).

Ruitenbeek, J. 1992. The rainforest supply price: A tool for evaluating rainforest conservation expenditures. *Ecological Economics* 1 (6): 57–78.

Rural Advancement Foundation International. 1994. Food patent challenge. In *Agenda for a Critical Decade: Strategic Plan 1993–2002.* Turrialba, Costa Rica: CATIE.

Sahlins, P. 1994. *Forest Rites: The War of the Demoiselles in Nineteenth Century France.* Cambridge: Harvard University Press.

Saragoza, Federico Mayor. 1995. Speech inaugurating City of Knowledge on April 22, 1995, Panama City.

Schmidt, R. 1993. International leadership. In *Environmental Leadership,* edited by J. K. Berry and J. C. Gordon. Washington, D.C.: Island.

Schmink, M., and C. H. Wood. 1992. *Contested Frontiers in Amazonia.* New York: Columbia University Press.

———, eds. 1987. The "political ecology" of Amazonia. In *Land at Risk in the Third World: Local Level Perspectives,* edited by P. D. Little and M. M. Horowitz. Boulder: Westview.

Schneider, R. 1994. Government and the economy on the Amazon frontier. Latin America and the Caribbean Technical Department, World Bank (Washington, D.C.), Report 34.

———. 1992. Brazil: An analysis of environmental problems in the Amazon. World Bank (Washington, D.C.), Report 9104-Br.

Schroeder, R. A. 1993. Shady practice: Gender and the political ecology of resource stabilization in Gambian garden/orchards. In *Economic Geography,* edited by W. W. Atwood et al. Worcester, Mass.: Clark University Press.

Schultz, T. W. 1979. Investing in people. Lecture given at Nobel Prize ceremonies. In *Agricultural Development in the Third World Countries.* 2d ed. Edited by C. K. Eicher and M. M. Staatz. Baltimore: Johns Hopkins University Press.

Schulze, W., G. McClelland, and J. Lazo. 1994. Methodological issues in using contingent valuation to measure non-use value. Paper presented at workshop, Using Contingent Valuation to Measure Non-market Values. United States Department of Energy and Environmental Protection Agency.

Southgate, D. 1994. Tropical deforestation and agricultural development in Latin America. In Brown and Pearce (1994).

Southgate, D., R. Sierra, and L. Brown. 1989. *The Causes of Tropical Deforestation in Ecuador: A Statistical Analysis.* London: London Environmental Economics Centre, International Institute for Environment and Development.

Southgate, D., and M. Whitaker. 1994. *Economic Progress and the Environment: One Developing Country's Policy Crisis.* Oxford: Oxford University Press.

Speth, J. G. 1993. With a soul and a vision: A new approach to development, and a new UNDP. Address to the UNDP staff, United Nations Secretariat.

Thrupp, L. A. 1989. Legitimizing local knowledge: From displacement to empowerment for Third World people. In *Indigenous Knowledge Systems: Implications for Agricultural and International Development,* edited by D. M. Warren. Ames: Iowa State University Press.

Toffler, A. 1990. *Power Shift: Knowledge, Wealth, and Violence at the Edge of the 21st Century.* New York: Bantam.

UNCED. 1992. *Agenda 21.* Rio de Janeiro: UNCED.

UNDP. 1994a. *Human Development Report.* New York: UNDP.

UNDP. 1994b. *Sustainable Human Development and Agriculture.* UNDP Guidebook Series. New York: UNDP.

West, P. C., and S. R. Brechin, eds. 1991. *Resident Peoples and National Parks: Social Dilemmas and Strategies in International Conservation.* Tucson: University of Arizona Press.

Winterbottom, R. 1990. *Taking Stock: The Tropical Forestry Action Plan After Five Years.* Washington, D.C.: World Resources Institute.

Wolters, O. W. 1967. *Early Indonesian Commerce.* Ithaca: Cornell University Press.

World Bank. 1992. *World Development Report.* Oxford: Oxford University Press.

Index

ACTS (African Centre for Technology Studies), 84–85
adaptive management, 12–16, 20, 21, 171, 172–73
aesthetics of forests, 2, 4
Africa, 2, 3–4, 42; subsistence farming in, xi, 5; deforestation in, xii, 25–26, 67, 96; poverty in, 16, 175; cropland vs. forestland in, 24; unrestricted migration in, 27; forest reserves in, 84; policy reform hindered in, 87–88, 89; forestry research in, 111; agricultural subsidies in, 139
African Centre for Technology Studies (ACTS), 84–85
age, forest rights and, 43
Agenda 21, 80, 81–82, 90, 94–95, 107, 108, 164
Agracetus/WRGrace, 116–17
agricultural subsidies, 138–39
agriculture, 22, 72, 88, 89, 181
agroforestry, 108–09, 131, 154, 155, 156, 160, 170
aid programs, 9–11, 74–77, 86–87, 122
Allen, J., 134
aluminum, 30
Amazon region, 53–55, 67, 133, 136, 143, 158

Anderson, D., 60
ASEAN (Association of Southeast Asian Nations), 111
Asia, 2, 3–4; subsistence farming in, xi, 5; deforestation in, xii, 25–26, 96; poverty in, 16, 175; cropland vs. forestland in, 24; closed forests in, 67; forestry research in, 111; closed frontier in, 127
Asian Development Bank, 70
Association of Southeast Asian Nations (ASEAN), 111
atmosphere, 13

Bagak Sahwa (Indonesia), 48–51
bamboo, 6
banana, 46
Bangladesh, 18, 24
Barnes, D., 134
Benbrook, Charles, 170
Berry, S., 57
Bihar (India), 93
biodiversity: res reps' views on, 2, 4; international initiatives on, 81, 83, 117; effect of sustainable agriculture on, 113; willingness to pay for, 141; "global bargains" for, 145, 180; forest strategies for, 152, 153